MATHEMATICAL
Imagining

MATHEMATICAL

Imagining

A ROUTINE FOR
SECONDARY CLASSROOMS

Christof Weber

Foreword by John Mason

Stenhouse
PUBLISHERS

PORTSMOUTH, NEW HAMPSHIRE

Stenhouse Publishers
www.stenhouse.com

Originally published as *Mathematische Vorstellungsübungen im Unterricht* by Christof Weber, copyright © 2010 by Friedrich Verlag GmbH. Friedrich is not responsible for the quality of the translation.

Library of Congress Cataloging-in-Publication Data

Names: Weber, Christof, author.
Title: Mathematical imagining : a routine for secondary classrooms /
    Christof Weber.
Identifiers: LCCN 2019036366 (print) | LCCN 2019036367 (ebook) | ISBN
    9781625312778 (paperback) | ISBN 9781625312785 (ebook)
Subjects: LCSH: Mathematics—Study and teaching (Secondary)
Classification: LCC QA11.2 .W43 2020 (print) | LCC QA11.2 (ebook) | DDC
    510.071/2—dc23
LC record available at https://lccn.loc.gov/2019036366
LC ebook record available at https://lccn.loc.gov/2019036367

Figure 6.4 © 2019, ProLitteris, Zurich. Foto: Peter Lauri.
Figure 6.5a–6.5d Courtesy Art Affairs, Amsterdam. Foto: Studio Brandwijk, Badehoevedorp.
Figures 7.3 and 7.4 Copyright John M. Sullivan, Technische Universität Berlin, Germany.

Cover design by Cindy Butler
Interior design and typesetting by Shawn Girsberger

Manufactured in the United States of America

PRINTED ON 30% PCW
RECYCLED PAPER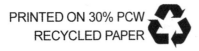

26   25   24   23   22   21   20   9   8   7   6   5   4   3   2   1

CONTENTS

FOREWORD

The power and importance of mental imagery has been recognized as long as people have sat around campfires and told stories. It is through stories evoking images that we experience vicarious emotions and the consequences of being ruled by them, and it is through forming mental images that human beings plan or prepare for future action. In the Upanishadic metaphor for human psyche as a horse-drawn chariot, mental imagery corresponds to the reins, the means by which the intellect communicates with and directs the horses, the emotional energy made available by the senses.

The case can be made that all mathematicians exploit imagery at least implicitly, and in some cases, there is evidence of exploiting it explicitly: for example, Begehr and Lenz (1998, 50) say about Jakob Steiner, the great eighteenth-century geometer:

> *Sometimes he closed the curtains in order to darken the classroom so that students could better follow and "see" his geometric constructions which he was describing using words and sometimes his hands.*

Philosophers and educationalists such as Alfred North Whitehead and Caleb Gattegno, among many others, have recognized that education is largely about supporting learners in bringing to the surface, developing, and using their natural powers. Of the many powers we all possess, the power to imagine—and to express what we imagine in words, diagrams, pictures; in manipulating material objects; in gestures, movements, and sound—is at once the most basic and possibly the most far reaching of them all.

Mathematics exploits and depends on this power to imagine. It is much more than "glorified arithmetic." It is about discerning details and recognizing relationships between those details and then extracting general properties that can be perceived as being instantiated in those relationships. Put another way, the power to imagine beyond the confines of the material world is both essential to mathematics and what gives mathematics its evident and undoubted power.

Even though some people appear to be unable to "see" pictures in their mind, they do *do* something in response to the invitation to imagine. So it may be helpful to allow the word *imagery* to refer to any and all forms of virtual-sensory experience. Furthermore, repeated exposure to invitations to imagine can develop the strength and power of imagery. In my own case, I was much stronger at detecting errors in strings of symbols, a form of algebraic sensitivity, than explicitly imagining mathematical objects, until I joined a group of people working on and with mental imagery. Over time, I found that I had developed a geometrical sense of configurations, which were the subject of my research, even though they were multidimensional and did not belong in or fit into Euclidean space. I still do not know whether I literally "see" mentally or whether I have a multifaceted "sense" of what I am thinking.

What is perfectly clear is that *not* making use of learners' powers to imagine and *not* supporting them in developing that power is like trying to teach someone to throw a ball with their hands tied behind their back. It makes no sense at all.

In this book, you will find not only some exercises through which to develop your power to imagine and to express what you imagine but also some very specific suggestions about how to do this in an effective way pedagogically, making use of the structure of human psyche. When learners express what they are imagining, completely unexpected aspects may emerge, and the book gives advice on how to exploit what students come up with. I myself have looked for ways to avoid or sideline mathematically inappropriate or unproductive images that emerge, but here you will find ways to encourage learners to modify or restructure those images so that they become productive. You will also find explicit proposals for how to develop your own imagery tasks to use with learners.

There is a common expression in English used in response to what someone is saying: "I see what you are saying." Some people advocate also using the phrase "I hear what you are saying," especially when the speaker is more audile than visile in their propensities. But if you really want learners to experience these, to be able to say, "I see what you are saying" and "I hear what you are saying," it is necessary to become expert at "saying what you are seeing" and "saying what you are hearing," in other words, to be articulate in expressing what you are imagining. To do this requires ongoing and repeated practice at imagining, at becoming aware that you are imagining, and at being aware of what it is that you are imagining. This awareness can then feed into becoming aware of how you express what you are imagining to others. The discipline offered in this book is likely to provide a firm foundation for this ongoing study.

To my mind, the greatest contribution this book can make is to bring the reader to awareness of their own imagery and awareness of how to speak *from*, not simply *about*, what they are imagining. Through this awareness, the reader may be able to make a significant educational contribution to the life of learners; for though these may or may not continue through life—imagining mathematical objects and relationships—their lives will undoubtedly be enhanced and enriched by becoming articulate in expressing, in whatever medium suits them, what they are imagining. Of course, to be maximally effective, this has to take place in a caring atmosphere, in which the teacher evidently cares for both learners and mathematics. Effective pedagogy resides in a balance between these foci of concern.

John Mason
Professor Emeritus, Open University
Honorary Research Fellow, University of Oxford

PREFACE TO THE ENGLISH EDITION

I believe that the image is the great instrument of instruction. What a child gets out of any subject presented to him is simply the images which he himself forms with regard to it. . . . I believe that much of the time and attention now given to the preparation and presentation of lessons might be more wisely and profitably expended in training the child's power of imagery.

—John Dewey, *My Pedagogic Creed*

The power to imagine is usefully called upon explicitly and can be developed with practice. If imagination is not called upon in mathematics, then a powerful link to the emotions is neglected, and motivation-interest may suffer. If expression in multiple forms is not encouraged, then learners may form the mistaken impression that mathematics does not offer opportunities for creativity. If learners encounter a very limited range of images, and a very limited range of expressions, they are likely to form the erroneous impression that mathematics is a very limited domain of human experience.

—Sue Johnston-Wilder and John Mason, *Developing Thinking in Geometry*

During my twenty-five years of teaching mathematics at the secondary-school level in Switzerland, I have seen firsthand the truth of these quotations—yes, we teachers can encourage our students to use their imaginations and mental images for learning, enjoying, and better understanding math. Mathematical Imagining was also the theme of my doctoral research, which helped to make explicit my tacit knowledge in action. In turn, the dissertation became the basis for this book, originally published in German in 2010 under the title *Mathematische Vorstellungsübungen im Unterricht*.

Since its publication, I have led several dozen professional development courses on using Mathematical Imagining in the classroom for teachers in German-speaking countries and also given many presentations about it. Meanwhile, interest in the topic has grown—and use of the Mathematical Imagining routine is spreading. So, it seems that my "child" has now matured enough to leave home, go out into the world, and meet teachers and students in other countries. To that end, I am pleased to be able to share the book with you now in this English version.

The term "exercises in imagining" was used previously by the mathematicians John Conway, Peter Doyle, Jane Gilman, and William Thurston. In their university summer workshop titled "Geometry and the Imagination," they promoted using imagery at the university math level, offering a series of exercises in Mathematical Imagining (Conway et al. 1991). But it is not only the occasional academic mathematician that considers imagery to be crucial for understanding. For example, some universities are now including work with imagining tasks in their mathematics teacher education programs. Furthermore, a number of preservice math teachers have incorporated Mathematical Imagining into their student teaching practice and written papers for seminars and bachelor's and master's theses that demonstrate the effects of using Mathematical Imagining in classrooms.

This handbook provides you with a collection of ideas and impulses to use in your lessons—ideas and lessons I have developed and refined in my own classroom, with my own students. It also demonstrates that imagining tasks can be implemented with certain nongeometrical content and can be used productively with your high school students so that they, too, can better learn, understand, and use the universal language of mathematics.

I wish you well in your use and exploration of the powerful and fascinating routine of Mathematical Imagining.

ACKNOWLEDGMENTS

My first thanks go to Stenhouse Publishers, namely, Tracy J. Zager and Dan Tobin. Their enthusiasm and patience contributed significantly to the publication of this book. Special thanks also to the production team, especially Shannon St. Peter, Jay Kilburn, and Stephanie Levy for their diligent and beautiful work.

I would also like to thank Gina Billy for all her effort and skill translating my German book into English, and Charles G. Gunn for clarifying many specific questions about mathematics and language that arose when we translated the exercises.

Furthermore, I would like to thank . . .

. . . all the high school students who let themselves be challenged by exercises in Mathematical Imagining

. . . all the student teachers who discussed and argued with me about the new routine

. . . all my teacher colleagues who wished for this book

. . . as well as the following people who have advised or supported me:

Alf Coles	Christian Rüede
Peter Gallin	Stephanie Thomas
Beste Güçler	Joy Titheridge
Martin Knecht	Jeff Weeks
John Mason	

. . . and—last but not least—

Andrea Zink.

Without her, this book would never have been.

AN INVITATION INTO MY CLASSROOM

[Mathematical Imaginings] get me interested,

because they are more vivid than normal math teaching.

—Diana, high school student

My high school students stream into class late in the afternoon, many coming from gym. They drop their bags next to their chairs and put their math materials on their tables. I switch off the lights and ask students to prepare for an exercise in Mathematical Imagining. There is a perceptible change in energy—students are familiar with this weekly routine, and they push their books, notes, and pencils out of their way and take a few focused breaths. I invite students to sit comfortably and close their eyes. Some students sit up, hands relaxed in their laps, with their eyes closed. Others choose to fold their arms on their tables and drop their heads into the spaces they've made, to further block out light and distractions. They rest their foreheads on their forearms and listen, expectantly. I pick up my written exercise and begin reading it out loud, slowly and clearly.

> Imagine you are walking on a grassy field. You see three long strips of cloth lying in front of you in the shape of a *large triangle*. Stand in the middle of one of the strips of cloth with your nose and toes pointing into the triangle *in front of* you and stretch your arms out sideways to the left and right: they form a line *above* one strip—that is, parallel to and above one side of the triangle. . . .

I watch as students imagine. They are relaxed, concentrating, each one creating a personal field and triangle. I notice that, when I told them to imagine stretching their arms out sideways, a few students made barely perceptible movements in their shoulders or fluttered their fingertips a tiny bit, their physical bodies echoing what their *imaginary bodies* are doing inside their minds. I continue.

> Now begin moving *sideways* along the strip of cloth, keeping your nose and toes pointing *into* the triangle as you *sidestep* in the direction of your right arm, placing one foot *next to* the other, until you reach a corner where two strips of cloth meet. . . .

I wonder whether students are moving sideways as I intended when I wrote the exercise or whether some students have mentally turned their bodies and walked toward the first vertex. I remind myself to explore this question in our discussions later.

> Your *right* arm now extends beyond the figure, and your *left* arm is above the side of the triangle that you have just sidestepped along. Rotate slowly about your body's axis with your arms firmly outstretched so that your left arm begins to point *into* the figure. Keep rotating your body until your left arm has swept out the corner and arrived *above* the next strip of cloth, and then stop rotating.

I am curious how students are experiencing this moment, this chance to consider an interior angle of a triangle as a *dynamic rotation,* rather than a static measurement. They'll have two more chances to think about these turns.

> Begin sidestepping along the triangle's side, *this time* in the direction of your *left* arm, placing one foot *next to* the other, until you reach the second corner. Your *left* arm now extends beyond the triangle, and your *right* arm is above the side of the triangle that you have just walked along. Rotate again slowly so that your right arm initially points into the triangle, sweeps out the corner, and arrives parallel to and *above* the next side. . . .

I take a brief moment to conjure up my own imagined triangle—the one I created when I planned this exercise and can now access in an instant. Taking this moment to recollect helps me connect to the mathematics and also helps me ensure students have sufficient time to imagine their triangles in detail. After this pause, I bring them back to the start.

> Continue sidestepping and rotating like this until you arrive back where you started from. . . .

I am delighted to hear some audible gasps as students return to the start. I ask my first questions.

> How are you standing now?
>
> What has happened?

I give students time to contemplate these questions individually, quietly. Asking students, "How are you standing now?" encourages them to compare which way they were facing when they started and which way they are facing when they ended up. I expect some students are mentally going around the triangle again, either deepening or modifying their original journey. Students who may have turned in different directions at the vertices, rather than making all their turns in the same direction, have their first chance to adjust their image and move around the triangle again. The question "What has happened?" encourages students to think about their mental journey around the triangle as a whole, perhaps even from a new, bird's-eye perspective.

When I sense the class is ready, I ask the question that I always ask at the close of this routine:

> What did you imagine during this exercise in Mathematical Imagining?

I wait, again, for students to consider this question, before inviting to them to "come back to class." Students open their eyes, eager to discuss their imaginings with their peers. I invite students to talk about their imagining in pairs or threes. I walk around, listening as students talk with one another about which way they were facing at various points along their routes. To help their classmates understand their points of view, some students spontaneously make quick sketches or use gestures. Students are eager to share and are curious about one another's imaginings, so they listen intently. I see a few students close their eyes and revise their triangles in response to what they've heard from their classmates.

After a few minutes, I ask students a new question.

> What is the sum of the interior angles of a triangle?

Students now have a chance to reason together about why they have made a half-turn from beginning to end. Why a half-turn, exactly? And is it exactly a half-turn? Will it always be a half-turn, regardless of the triangle? These questions help lay the mathematical foundation for the interior angles theorem.

When conversation naturally begins to trail off, I ask students to take out their journals. Sometimes, I close our imaginings without journaling, but in this case, I decide to make time for students to capture their imagining before it starts to fade because their imaginings will be foundational to our upcoming work. For example, we'll eventually want to consider what happens when shuffling around a flat polygon with four vertices? Five? n? That discussion will be much richer if students can readily access their personal, mental triangles and modify them by adding sides and vertices. Therefore, I ask students to take a few minutes to respond to these prompts in writing before moving on to the rest of my lesson:

> Record the mental images—both pictures and actions—that you imagined during this exercise.

> Which of these images and actions were useful to you? Which got in your way? What revisions did you make?

Later, I'll look closely at their writing and sketches. I might select a few examples to share and have students discuss tomorrow. Or, I might choose to have all students leave their open journals on their tables and then walk around to read and comment on other students' work. Or, I might read them solely to inform my planning for upcoming lessons. Either way, this Mathematical Imagining gives my students a chance to create mathematics for themselves, and it helps me to see what is endlessly fascinating but not so easy to see: how my students think mathematically.

INTRODUCTION

An imagining task is a completely different world from a worksheet.

I can move there more freely.

—Peter, high school student

EXERCISES IN MATHEMATICAL IMAGINING

Mathematics can be an intellectual adventure, an adventure that takes place in our heads. It then thrives on the ability to imagine—objects and information are pictured in the "mind's eye" and further processed mentally. How can this imaginative aspect of mathematics be made to come alive and be implemented in the classroom?

Exercises in Mathematical Imagining cultivate the imaginative dimension of our discipline. One of the strengths of this new routine is that students engage with mathematical content directly and in a very personal way. This means that all students, not only the "high achievers," find a personal, inquiry-based approach to math. They learn to visualize math exercises more vividly and to work with their individual mental images. Students then find it easier to develop their own ways of solving a task. As we started to see in the introductory triangle example, working with imagining tasks in the classroom also gives you insight into your students' thought processes and how they go about tackling problems. Let's think about this abridged example of a different imagining task:

> Imagine a plastic cup lying on the floor. Give the cup a nudge so that it begins to roll. What does the path it takes look like?

This short text prompts your imagination, especially if instead of reading it out loud, you put yourself in the role of the listener. The words evoke mental images and actions that depict and elaborate the situation described. Thus, when you give the cup a mental push, you can "see" and perhaps even "sense" how it rolls forward. If your cup looks like a normal cup—a bit wider at the top than it is at the bottom— it will roll in a curve rather than in a straight line. You might have to push the cup more than once before it completes its trajectory and is finally back where it began. Is the trajectory in fact circular? With this conjecture, you are being mathematically active, and you are right in the middle of your own mathematical world.

This is precisely the effect of imagining tasks in the classroom. They open up a space in which students can engage with a mathematical situation, develop their own ideas and mental images, pose mathematical questions—and thus do mathematics in the best possible sense. Math begins in our heads!

INDIVIDUAL MENTAL IMAGERY IN THE CLASSROOM AND ITS VALUE FOR THE LEARNING PROCESS

The previous example encourages us to imagine a certain situation. However, it deliberately gives only a few pointers in order to leave room for individual, yet math-related mental images to take shape. Neither the exact shape nor the size and diameter of the cup are described, even if the wording *plastic cup* suggests a truncated right circular cone that can be held comfortably in one hand. The gradient and composition of the floor surface and the speed at which the cup rolls likewise remain unspecified. Only by exploring, elaborating, and thus acquiring an increasingly clear picture of a mathematical situation does it become mathematically accessible.

Once this foundation is present, it is possible for your students to perform actions and experiments in their minds, to formulate conjectures, and to be mathematically active. How they answer the question about "what the path looks like" (the shape of the trajectory) depends on how they imagine the rolling cup. Previous experience and basic mathematical knowledge allow them to make simple conjectures, such as "The cup rolls in a curve" or "The form taken by the trajectory depends on the shape and mass distribution of the cup." Whether it is a closed or even a circular "curve" is not yet clear. And neither is it clear which of the cup's geometrical and physical characteristics determine the exact form of the trajectory.

Other questions follow: Can two cups with different shapes follow the same trajectory? What does a solid figure that rolls along a helical trajectory look like? However, the students ask themselves these questions only after they have imagined the situation.

Exercises in Mathematical Imagining can prompt productive lines of thought; indeed, the mental images themselves can be enormously productive. For instance, instead of the plastic cup, we could imagine a single-axis vehicle with two wheels of different sizes. Or the plastic cup might be lengthened into a full cone by the addition of an appendage extending beyond the cup's truncated cone. Students construct these kinds of mental images themselves, and because they contain the nucleus of a possible answer, they help to establish a solid mathematical basis. Despite this, Mathematical Imagining goes largely unheeded in math teaching. In short, there is untold potential here just waiting to be harnessed.

EXERCISES IN MATHEMATICAL IMAGINING: MY OWN EXPERIENCE AS A TEACHER

I was pleasantly surprised by even my very first experiments with exercises in Mathematical Imagining. That was because my high school students were unexpectedly open and willing to engage with them. And even after students had been working with imagining tasks over the course of a few school years, they continued to enjoy them very much—and kept asking me for more.

Over time, I have gained the impression that more students are fully alert and actively engaged during imagining tasks than is the case in the remainder of the lesson. This is particularly true of "lower-achieving" students. When sharing and comparing their mental images, these students are more animated and engrossed in the subject matter than I would have thought possible. It is precisely because imagining tasks focus on the personal imagery of each individual that math suddenly becomes relevant to everyone. Students feel personally involved, and they become increasingly willing to engage with mathematical questions.

For me as a teacher, there is another crucial benefit of this routine. Working with imagining tasks allows me to gain insight about what is going on in my students' heads when they are doing math. When I listen to my learners as they develop their own thoughts and ways of solving the task at hand, I often discover an enormous wealth of ideas. On closer scrutiny, there are always (in addition to the false steps) a number of productive approaches, and many of them run counter to my own way of thinking. In short, what is actually going on in my students' minds is often quite different from what I had expected or imagined it would be beforehand.

Using imagining tasks in your math lessons gives you greater insight into the way your students approach math problems and how they think mathematically. You can translate into action the old call for teachers to meet learners where they are, and you will find yourself surprised and delighted by what your students imagine mathematically.

HOW YOU CAN READ AND WORK WITH THIS BOOK

This book is, above all, written for application in the real classroom setting. It therefore includes an extensive collection of examples. In Part 2, you will find more than thirty exercises in Mathematical Imagining, including the preknowledge and skills required for students to do the task. They come from all areas of school mathematics and have been used with grades nine to twelve. Simpler imagining tasks are also suited for use at middle school with grades seven and eight. Within the exercises, I have provided mathematical and pedagogical commentary to help you plan and implement the exercises yourself. These notes are based on my experience using these exercises in my own classroom and are designed to help you anticipate what images the exercises might evoke, facilitate discussions among your students, plan follow-up questions and extensions, and so on. If you are primarily interested in the practical aspects of this new routine, you might want to begin by browsing Part 2.

I also hope to inspire you to use exercises in Mathematical Imagining in your lessons and to develop your own tasks. You will find background information and guidelines on how to do this in Part 1. Here you will learn what does—and does not—constitute an exercise in Mathematical Imagining, what the tasks hope to achieve, and the results they can bring forth (Chapter 1). You will also find notes on how to implement the Mathematical Imagining routine in your classroom, plus how students' individual mental images can be used in the learning process (Chapter 2). The final section of Part 1 identifies criteria you can use to develop your own exercises in Mathematical Imagining (Chapter 3). If you want to understand the purpose and possibilities of this routine, then I recommend beginning with Part 1.

Wherever you choose to start, I hope that this book will "whet your appetite" for working with imagining tasks and provide you with some new ideas and inspiration to enrich your lessons and provide your students regular opportunities to "see" math.

_ P A R T   1 _

The Mathematical Imagining Routine: Background and Guidelines

The real problem which confronts mathematics teaching is not that of rigor,
but the problem of the development of "meaning,"
of the "existence" of mathematical objects.

—René Thom, "Modern Mathematics: Does It Exist?"

CHAPTER

1

What Is the Mathematical Imagining Routine?

The examples of the triangle in the field and the rolling plastic cup described in the introduction gave you glimpses of Mathematical Imagining. Implementing this routine in the classroom is quite a new teaching idea—and people often associate imagining with many different things. That is why we'll begin by trying to clear up a few misunderstandings. Most important, though, in this chapter, I aim to make clear the purpose of the Mathematical Imagining routine, what it can accomplish, and why using it in the classroom is so worthwhile.

WHAT THE MATHEMATICAL IMAGINING ROUTINE IS—AND IS NOT

Perhaps the most concise definition of the Mathematical Imagining routine follows:

The Mathematical Imagining routine includes rich math tasks students work on mentally, based entirely on their own mental images.

A more detailed characterization could go something like this:

The Mathematical Imagining routine describes simple but rich tasks. The math involved is situated in the context of real objects and activities. The wording of an imagining task invites the listener to imagine this situation and explore it in their mind. The listener draws on, develops, and hones the individual mental images that arise in order to answer a given mathematical question. This exercise takes place without using objects, notes, or other external aids.

This description reminds us of a number of known approaches, such as mental geometry and guided imagery, and yet it would be a mistake to equate Mathematical Imagining with either of these. Now, why is that?

Misunderstanding no. 1: Mathematical Imagining is mental geometry for spatial visualization training

Mental geometry is geometry performed without any representational aids in the form of sketches or physical objects. The task invites a person to see and manipulate a geometric shape in their imagination in order to solve a mathematical problem: it is geometry in the mind's eye. Like mental arithmetic, mental geometry problems are usually designed to train basic skills and techniques. The suggestion by the proponents of mental geometry is that repeatedly solving difficult geometry problems mentally trains spatial visualization skills. When this training aspect takes center stage, it is often at the expense of the mathematical content. Mental geometry tasks are then reduced to "brain training" and bring to mind certain questions on "intelligence tests." Creators of such tests include spatial visualization ability as one intelligence component. So, it comes as no surprise that, with the appropriate mental geometry training, individuals will do better on an "intelligence test" and achieve higher IQ results.

This result does not, however, mean that they have greater mathematical knowledge, skill, or understanding. Empirical research on teaching and learning has shown that knowledge and ability are much more clearly limited to a specific area than has long been assumed. For example, people who solve Sudoku problems or memorize poems on a daily basis are not training their logical reasoning or general memory but rather, their specific ability to solve Sudoku problems or learn poems. Skills and knowledge are not necessarily or easily transferred to other areas, as much as we may wish this would happen.

Exercises in Mathematical Imagining do have some similarities to spatial visualization tasks; they are likewise geared toward answering mathematical questions without using objects or notes. Nevertheless, they differ from mental geometry tasks in several respects. First, Mathematical Imagining may also involve nongeometric content. As long as the content can be visualized, it can form the basis of an exercise in Mathematical Imagining, including from the fields of arithmetic, algebra, or calculus. Second, many exercises in Mathematical Imagining also provide learners with material they can use to solve future mathematical problems. Imagining tasks encourage learners to explore and construct mathematical objects, to develop arguments, and to reason plausibly. With these heuristic processes, exercises in Mathematical Imagining go beyond finding solutions to challenging problems. Which brings us to the third and crucial way in which they differ from mental geometry tasks: within the Mathematical Imagining routine, you can directly address your students' individual mental images—the actual images they generate in their minds—and develop them further. We'll explore the power of this potential in Chapter 2.

Of course, the ability to think spatially also plays a certain role in exercises in Mathematical Imagining. However, because these tasks place mathematical content in the context of real objects and activities, basic math competencies and knowledge of the subject matter are just as relevant here as the ability to visualize. Furthermore, the mathematics in Mathematical Imagining tasks is not merely a means to the end of training a "muscle," as there is no guarantee that it will make someone fit to solve other math problems. What happens, though, is students acquire new knowledge because they are building on their preexisting network of specific mathematical knowledge.

Misunderstanding no. 2: Exercises in Mathematical Imagining are guided imagery activities

Guided imagery activates an individual's imagination and appeals to listeners' emotions. It can be used in the classroom setting to combine intensity and relaxation in order to lead learners into a state of relaxed attention and to facilitate their learning processes. Sometimes, guided imagery is also more structured and provides a given theme, such as "Imagine you are sitting inside a flower. . . ."

Exercises in Mathematical Imagining are likewise designed to evoke and focus on students' mental images. The aim, however, is not to increase concentration—although this is a side effect—but, rather, to use individual mental images to stimulate independent mathematical thinking. Imagining tasks draw learners' attention to the potentially productive and counterproductive effects their mental images can have on their mathematical reasoning. In contrast to guided imagery, exercises in Mathematical Imagining place greater limits on the imagination because they address only math-related mental images and disregard those that are associative or decorative in nature. This means that more guidance is sometimes needed in imagining tasks.

To sum up, there are areas of overlap between Mathematical Imagining and mental geometry or guided imagery. However, the fact that exercises in Mathematical Imagining are geared toward concrete mathematical content and processes means that they go beyond the related approaches and acquire a specific dimension of their own.

WHAT THE MATHEMATICAL IMAGINING ROUTINE IS DESIGNED TO ACHIEVE

The Mathematical Imagining routine seeks to bring interesting mathematical content into the classroom. However, these tasks are remarkable not so much because of the teacher's special selection of material or innovative lesson preparation but because these exercises focus directly on the learners' subject-specific, individual, and math-related mental images. Now, what changes does this method intend to effect?

We have all often heard (and perhaps made) the complaint that high school math lessons focus too heavily on predetermined solutions and procedures. And yet we as teachers also know how difficult it is to structure a lesson that promotes active, inquiry-based learning. So, for simplicity's sake, we mostly just go ahead and do the active part ourselves. We prepare the "enlightening, informative introduction" to (hopefully) awaken our students' interest or the "easy" explanation at home to save them from making any potential false steps. The classroom result is that we teachers give an expert demonstration, as in "Look here! I know how it works, and I'm showing you how to do it." Unfortunately, this implicit pride on the part of the "expert" who already "knows it all" leads to learners not really actively engaging—and, at best, being motivated to just imitate the teacher. In contrast, the Mathematical Imagining routine, like other teaching tools and approaches, aims to help students become active learners who are more empowered and involved in constructing, understanding, and retaining their own knowledge.

Heuristic and mathematical processes

The issues raised here are not new. The German-born Dutch mathematician and educator Hans Freudenthal, for instance, called for "mathematics [to be] taught not as a created subject, but as a subject to be created." He therefore proposed writing "a textbook dealing with a particular mathematical domain on two levels: left and right the same subject matter; left such as is grasped by one's own first learning process, and right such as is formalized after it has been grasped" (Freudenthal 1978, 71).

The Hungarian mathematician and educator George Pólya argued similarly. He held the view that unlike mathematics itself, mathematics education had to be interested not only in the solution of a mathematical problem but just as much, if not more so, in how this solution evolved or, as he described it, in the "progress of the solution" (Pólya 1962, ii, 1–12). He analyzed the genesis of solutions in order to understand how they had come to be. In this way, Pólya developed *heuristic processes*, or approaches that allow learners to discover solutions for themselves through plausible and provisional thinking, which can be used strategically and as a guide to solving mathematical problems (Pólya 1945, 1981).

The Mathematical Imagining routine, first of all, creates a space in which different paths to a solution or solutions can evolve. In this respect, they operate on the left side of Freudenthal's textbook. They encourage students to engage with a problem and visualize it in order to ultimately have the related mathematical content at their disposal. This initial engagement and sense-making forms the basis upon which students further explore the content. They subject it to mental experiments, raise questions, even question the content itself, and thus, may make new connections. As a consequence, students formulate hypotheses or reach plausible conclusions. (You will see an example of how this process plays out in the classroom in Chapter 2.) All these activities are, according to Pólya, heuristic in nature (Table 1.1, column on the *left*). Only after students have come up with their own unpolished thoughts and mental images do they begin to examine, structure, and process. This is when students engage in classical mathematical thinking, such as modeling, proving, and generalizing, and there is a shift to the right side of Freudenthal's "textbook" (Table 1.1, column on the *right*).

Learning processes that fail to incorporate individual thoughts and mental images are incomplete. In such "cutoff" processes, students then tend to merely imitate the teacher, and understanding remains superficial. Exercises in Mathematical Imagining intentionally operate in an area that goes beyond modeling,

TABLE 1.1   Mathematical Imagining tasks encourage heuristic and mathematical processes

Heuristic processes	**Mathematical processes**
• Engaging and visualizing	• Representing and modeling
• Making content available and exploring	• Mathematical reasoning and proving
• Experimenting, guessing, plausible reasoning, and questioning	• Varying and generalizing
Objective: To produce solutions and knowledge	*Objective: To validate solutions and knowledge*

reasoning, and communicating—the necessary but partial "standards for mathematical practice" of standards and curriculum. Because exercises in Mathematical Imagining specifically trigger heuristic processes, they also invest mathematical processes with another quality: mathematics is anchored in the individual and is consequently more understandable and invested with meaning-making for each student personally. With Mathematical Imagining tasks, you can structure lessons in which both heuristic and mathematical processes take place.

Going beyond right and wrong: **Why we should work with individual mental images**

In the context of Mathematical Imagining, our focus is not only on images that are immediately recognized as being math related but also on images that can be *made* meaningful for mathematics. This is complex work partly because the mental images students construct include both the actual pictures they see with their mind's eyes (e.g., in the triangle exercise from the invitation on page xii, the strips of cloth forming a triangle on the grass) and also the mental actions they carry out (sidestepping along the triangle, rotating at the vertex). This is due to the dual nature of imagining itself. *Mental pictures* are static in nature, whereas *mental actions* add movement and dynamics. This duality is partly because images are usually taken to be visual, and indeed, sight is the dominant sense for most of us. But there are various ways in which people have a "feel" for mathematical concepts, which may invoke other senses, such as auditory or tactile sensations, or a motor sense of "movement."

Imagining also evokes individual, idiosyncratic mental images. For example, in the triangle exercise, we might imagine different strips of cloth or different triangles. *Individual mental images* like these can support meaning for mathematics. For instance, our number line represents a mathematically correct interpretation of the real numbers. This geometric interpretation is the mathematical standard, and

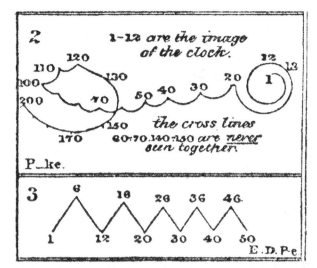

FIGURE 1.1 Individual mental images of the set of real numbers (Galton 1883)

only with it are statements such as "π is close to 3" or "a sequence approaches zero" meaningful through their reference to the geometric concept of distance.

However, when you ask different people how they arrange the set of (real) numbers in their own minds, their individual images of the number line are often quite astonishing. The number line does not play a role in most people's mental images of numbers. Instead, they arrange the numbers in nonequidistant, spiral-shaped, or zigzag patterns, as shown in Figure 1.1. Such images are not simply reproductions of sensory perceptions but are in fact heavily influenced by personal experience and previously constructed knowledge (see the clock in the upper part of Figure 1.1). These individual pictures supplement the standard ones, serve as a guide when making calculations, and reflect individuals' personal, unique ways of looking at mathematical content.

In individual mental images, a mathematical content area (here the set of real numbers) is combined with a geometric interpretation (the number line). Although they need not be correct from a mathematical point of view, they are often quite useful and can support mathematical thinking and reasoning. For instance, the size of two numbers can be compared on the basis of the two individual mental images pictured.

The focus on individual mental images in the classroom may call to mind literature classes where teachers regularly ask students to give their personal point of view. This applies as well in many typical math lessons in which students may be asked to picture something in their minds. The mental image they come up with is then used for further mental work ("Imagine a cube. Tilt it so that. . . ."). However, for the most part, teachers are interested only in those student images that align with accepted mathematical knowledge. The many remaining images students have envisioned go largely unheeded.

To further complicate matters, students often express their mental images in language that is awkward and sometimes even mathematically incorrect. When this happens, students are speaking, as the German physicist and educator Martin Wagenschein put it, "the language of trying to understand" and not "the language of having understood" (Wagenschein 1970, 162; author's translation). As a consequence, teachers often find their students' individual mental images pointless or bothersome, rarely take them up in class, and may equate them with misconceptions.

What opportunities are we missing here? People have the genuine ability to imagine something and develop their mental images, or, as the British mathematics educator John Mason put it, "these natural and far-reaching powers [are] possessed by all learners who come to class" (2002, 81). Artists, scientists, and scholars consider mental images fundamental to their work. Further, each person generates and controls their own individual mental images and also the way they imagine things. This means each of us is personally connected to our own mental images, and we experience ourselves as the originators and manipulators of what we imagine. To begin with, we are inevitably alone with our mental images. They cannot be seen from the outside, and they generally differ from the imaginary worlds of others. Therefore, students do not experience the concluding question of the Mathematical Imagining routine—"What did you imagine during this exercise in Mathematical Imagining?"—as a teaching trick or purely rhetorical question. Rather, the question triggers genuine interest and a real sense of personal involvement because it is addressed to each student: How did *you* imagine it? When students think about their own mental images, they become more willing to engage with heuristic processes—and hence with mathematics (Table 1.1). Moreover, since the question about individual mental images considers aspects that do not fall within approved mathematical knowledge, math lessons become much more open to unpolished thoughts and images, personal struggles, and trial and error, as we'll see in the examples throughout the book.

Individual mental images reveal not only a student's personal conceptions but also their attributions of meaning, that is, the rules and concepts students have adopted. Therefore, mental images, such as the personal number line, have an explanatory power that we should not underestimate. However, as productive as this power can be, rules students have ready, such as *When I see a minus sign in front of a number, I know it's negative,* can be counterproductive. When such dominant concepts are not taken up and worked on in the classroom, they remain intact. They reappear in stressful situations, such as taking exams, causing students to "forget" what they have learned and preventing them from working confidently and accurately. Counterproductive conceptions are so influential precisely because they have not been addressed and elaborated on in the class. We'll focus on an alternative teaching approach that takes up counterproductive mental images—rather than casts them aside—in Chapter 2.

The Mathematical Imagining routine thus pursues two fundamental goals:

1. *Learning objective*: Mathematical Imagining seeks to promote independent thinking and hence inquiry-based learning. Students are encouraged to engage with a variety of different heuristic processes that can lead into different math activities.

2. *Teaching objective*: Mathematical Imagining seeks to focus on individual mental imagery, which provides insight into how learners approach mathematical questions. Teachers can take up these approaches and structure lessons in which students follow their own paths and work together to acquire new knowledge.

In Chapter 2, we will see how this can be done.

TO WHAT EFFECT? RESULTS FROM TEACHING THE MATHEMATICAL IMAGINING ROUTINE

In my teaching, I've found that exercises in Mathematical Imagining are worthwhile for a number of reasons. The immediate benefit of doing them is a practical one: in Mathematical Imagining tasks, students are actually working on the subject matter you are teaching them. Another immediate benefit is the increase in students' concentration and interest: students focus their attention on their own mental images and actions and the associated mathematical questions. I've found this to be true even when I used the routine in a class scheduled at a "bad" time, such as when my students were already showing signs of exhaustion from the school day or week. As one of my students put it, "I look forward to the imagining task on Friday at the beginning of the lesson. That's why I often find math more interesting on Fridays than on other days."

Hereafter, I want to share and describe four positive effects of regularly incorporating the Mathematical Imagining routine into your teaching practice. The first two show how learners use their genuine imaginative abilities for the sake of mathematics. The last two relate more to general pedagogy.

Effect no. 1: Generating mathematically productive mental images and actions

This book differentiates between different types of exercises in Mathematical Imagining. Two types, *construction* and *reasoning*, provide concrete help with the visualization of complex mathematical objects (such as an icosahedron) or content (such as the sum of the internal angles of a triangle). Other exercises in Mathematical Imagining of the types *problem-solving* and *paradox* encourage students to develop their own mental images in order to solve a mathematical question on the basis of imagery and not by means of calculation or formal reasoning.

In all instances, students generate their own *productive mental images* that take place with a high degree of inner involvement. As a rule, this means that students remember the mathematical subject matter of an imagining exercise longer than other content. It is also possible that they are able to sketch geometrical shapes more precisely following an imagining task than they can after they have actually been shown the shape for real. After an imagining exercise, students have new mathematical content at their disposal.

This new, available knowledge is always accompanied by a sense of meaning-making and improved understanding. In particular, exercises in Mathematical Imagining of the reasoning type help students gain a more thorough understanding of mathematical content. In this way, mathematical content (such as the sum of the internal angles of a triangle) becomes more accessible, and students are thoroughly acquainted with its properties and connections. As one student expressed it, "Exercises in Mathematical Imagining are real 'aha!' moments." The other types of exercises in Mathematical Imagining also result in greater meaning-making and understanding, namely, when students construct their own productive mental images or recognize how productive the mental images of their classmates are.

When the routine is conducted regularly, there are three further effects that have an impact beyond individual exercises in Mathematical Imagining.

Effect no. 2: Leveraging mental images and actions as a strategy for solving mathematical problems

We take little notice of mental images and their mental manipulation in our day-to-day lives, and they are rarely consciously and explicitly put to use for inquiry processes. In math lessons, too, teachers rarely mention, let alone make tangible, the benefits of imagining a topic and experimenting with it independent of the topic's reality or graphic representation.

In contrast, the Mathematical Imagining routine explicitly teaches students to use mental images and mental actions to explore and experiment with specific mathematical questions. It is a strategy that can be used to clarify mathematical questions. This strategy is acquired not through isolated attempts but accumulates through a long-term, "getting used to it," process. Only when students experience exercises in Mathematical Imagining on a regular basis can they incorporate this strategy into their existing repertoire. Students then have an additional, extremely useful alternative approach to working on mathematical questions. Time and again, students have described to me how helpful it is for them to begin by imagining a given mathematical situation in their mind's eye—as opposed to the usual approach of acting and calculating straight away. They find imagining particularly profitable for problem-solving tasks. After two years of lessons using Mathematical Imagining tasks, a student told me, "I now try to imagine math tasks more vividly, and this helps me to solve them more easily." Mathematical Imagining thus shifts from being a teacher-led educational tool, or routine, to a strategy that students can internalize and use on their own.

Effect no. 3: Promoting thoughtfulness and reflection in math classes

Reflection is sometimes included as being one of the pedagogical aims of school education. In this context, reflection means that when someone finishes school, they should not only know and be able to apply the subject matter but also be in a position to reflect on its meaning and purpose. The case is often made that in addition to *reflection on meaning and purpose*, teaching should also encourage *self-reflection*. Empirical research on teaching and learning supports this practice. Self-reflection, or metacognition in particular (a reflection on and control of one's own cognitive learning processes), fosters effective learning if and when such reflection takes place with regard to specific content and situations. One methodological approach for thoughtful instruction is the genuine sense of personal involvement expressed, for example, in the question, "What does this have to do with me?"

This personal reflection is exactly where exercises in Mathematical Imagining come in. The question, "What did you imagine during this exercise in Mathematical Imagining?" stimulates each student to self-reflect on their own cognitive processes in relation to mathematics: *How did my imagination process this imagining exercise? Where doesn't it work? What mental images and actions did I picture? Is it more productive to turn an object I have imagined or to move around it? What mental images get in the way?* This reflection is by no means limited to a student's own mental imagery and ways of imagining things. The fact that students are affected positively by exercises in Mathematical Imagining means they have enormous interest in sharing what they imagined—and hearing about their classmates' mental images. *How do my own mental images compare with those of other students in my class? How do they compare with mathematical models and concepts? Which of my peers' mental images could be productive for me?*

Through this regular self-reflection, the learners who otherwise tend to avoid or even refuse to reflect on mathematical subject matter gradually begin to engage in mathematical reflection. The "paradox"-type tasks appear to be particularly well suited to these students. As well as initiating reflection on one's own and others' ways of imagining things, these exercises are directly aimed at thinking about mathematical content. (An example is P1, which involves approaching the topic of infinity through mental images.) As one student said, "Some exercises in Mathematical Imagining have me thinking about them all day."

Effect no. 4: Communicating a more open and vibrant view of mathematics

The mathematics teaching we receive at school shapes our view of math. Students are generally presented with content as a *fait accompli*. They are expected to remember definitions and apply formulas and procedures. The result is that most learners develop a relatively narrow and unrealistic view of mathematics. Their math lacks an inchoate or experimental side and certainly has no place for creativity. To them, mathematics seems to be a "given." Every posed question has just one answer; open or ambiguous questions are never posed. And all of the answers are already known before students even try to find a solution. Moreover, learners have to conform to mathematics as embodied in the authority of their teachers. It is hardly surprising, then, that students are highly unwilling to engage with math and that they develop a negative view of it.

Mathematical Imagining tasks break down this narrow, static view by taking account of individual mental images and focusing on a kind of "math in the making." Learners can start to check out a potentially challenging concept through first exploring how their own piece of mathematics evolves through their own mental actions. They are called upon to formulate and communicate their own unfinished ways of imagining content before these conceptualizations are assessed according to mathematical standards. In this way, students experience firsthand that they can contribute to solving and developing mathematical problems. Imagining tasks thus emphasize that math's nature is process oriented and that its genesis depended—and depends—on the creativity and imagination of individuals. Exercises in Mathematical Imagining underpin a teaching approach that is committed to the communication of a more open and vibrant view of mathematics. To quote another student of mine, "You get a different view of math than you do with usual math tasks."

All in all, the Mathematical Imagining routine is a valuable addition to the methodological repertoire available to teachers. Now that we've discussed a bit about *why* you might want to incorporate Mathematical Imagining tasks in your classroom, it's time to focus on *how* and *when* to go about implementing this routine.

C H A P T E R

2

Implementing the Mathematical Imagining Routine

So, what do you need to know in order to successfully introduce and carry out imagining tasks in your lessons? What factors do you need to consider? In this chapter, I'll begin by giving you some general thoughts on organizing imagining tasks for use with your students. Then I'll share some different options for implementing exercises in Mathematical Imagining, so you'll be able to tailor this routine to your needs, teaching context, and varying constraints. In the last two sections, I'll share "keys for success"—lessons I've learned through teaching imagining tasks with my own students. I'll help you anticipate challenges and difficulties that might come up when you implement this routine, and I'll suggest possible ways to handle these issues.

WORKING WITH IMAGINING TASKS

Exercises in Mathematical Imagining are a vivid and effective supplement to other teaching tools, such as investigations or student presentations. In general, the following applies when facilitating Mathematical Imaginings:

- You need to schedule at least fifteen minutes for an imagining task in your lesson plan. Should the task be used to trigger a learning process that will carry over into several lessons, allow for about ten minutes more. (We'll discuss these options next.)

- So that a class can get used to this new teaching tool as quickly as possible, it can be advantageous to schedule imagining tasks for a particular class on the same day and period of each school week. The tasks also work well when implemented at "bad" times in your students' schedules, like right after lunch or gym or during the last lesson of the day. This is because especially when students are demonstrating signs of physical or mental fatigue, imagining tasks smooth the way—or even make it possible—for them to turn their attention to math.

- To obtain effects that go beyond a single exercise in Mathematical Imagining (pages 59–223), you will need to schedule tasks as regularly as possible. My personal experience has been that the ideal frequency ranges from once per week to once every two weeks.

- Students work completely on their own during some phases of imagining tasks and may not use any type of visual aids, such as notes, sketches, or objects. This is a quite demanding framework, as it requires total concentration from the entire class. At the same time, though, exercises in Mathematical Imagining also foster concentration. For these reasons, the tasks work especially well at the beginning of a class, but, of course, you can also work with them during other parts of a lesson.

- You do not assess student performance on imagining tasks, especially when it comes to a task having a right or a wrong answer. By removing the performance pressure, you send your students a credible signal that exercises in Mathematical Imagining are not just one more way of testing them. This "no grading" stance clearly demonstrates that the purposes of imagining tasks are to let your students take their own voyage of discovery and provide them with a learning opportunity. Leaving out grades is a means that greatly promotes open discussion about students' individual ways of imagining. It also encourages students to share their *real* experiences with the task—instead of saying what answer they think you are looking for or even trying to get you to tell them the "right" solution.

You can use most of the exercises in Mathematical Imagining in this book to either introduce a new topic or to revise and/or deepen an area that has already been covered:

- Each imagining task invites students to envision and explore specific mathematical content, which is why this teaching tool is particularly suited for introducing new material. Sometimes beginning with an imagining task can be fruitful even when the new content deals with things like learning a new term, sketching or making calculations, or engaging with new geometric concepts. The Mathematical Imagining routine invites all students to access the new content and stimulates their curiosity and intuition about the unfamiliar mathematics to come.

- Exercises in Mathematical Imagining also invite students to continually visualize specific content so that they can keep expanding and developing it. That is why you can also use imagining tasks retrospectively, for example, to shed more light on or to delve deeper into aspects of a familiar mathematical idea or term and give students opportunities to synthesize ideas and draw connections among them. It can be particularly productive to ask students to revisit an initial Mathematical Imagining and revise it based on their new understandings.

EXERCISES IN MATHEMATICAL IMAGINING IN CLASS: A MINIMAL AND AN EXPANDED VERSION

So how exactly can you work with imagining tasks in your classroom? I'll frame two possible ways of going about it. The first is a simple, minimal version you can integrate into your usual lessons without having to make major changes to the way you already structure your classes. In it, your students' individual mental images are activated and, in a sense, coaxed out of their conscious minds. These images then become the subject of a class discussion.

In the expanded version, the individual mental images are not just available for discussion but also developed further. In this way, the mental images form the basis of knowledge that has been worked out collaboratively. Later in this chapter, a concrete example lesson demonstrates the far-reaching impact this kind of math learning experience can have.

Please keep in mind that the following lesson versions are suggestions. Your own expertise, experience, and sensitivity will inform how you structure the work with exercises in Mathematical Imagining with your students. Also, it's important to note that regular use of the minimal version produces positive effects. The expanded version shows you the potential in imagining tasks. By sharing both, I hope to inspire you to experiment and to explore and discover your way to teach Mathematical Imagining.

Minimal version: Developing and discussing individual mental images

Every exercise in Mathematical Imagining runs through at least two phases: one of imagining that takes about five minutes and the discussion round of ten to fifteen minutes. Figure 2.1 shows the phases.

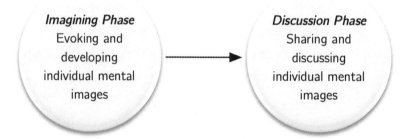

FIGURE 2.1   Phases of an imagining task (minimal version)

The imagining phase: Evoking individual mental images

Your students are all sitting in the classroom. You introduce the exercise in Mathematical Imagining by asking them to clear their desks of all objects and to sit comfortably. Tell them that, if they want to, they can use an arm to support their head, and invite them to close their eyes. During their very first imagining task, it's likely that not all your students will close their eyes, but after doing a few more, they may accept your invitation.

When introducing an imagining task, it helps to use wording similar to that used for introductions to concentration exercises or guided imagery. Here is an example, and in it, the ellipses (. . .) signify you should make a brief pause while speaking.

> Now, everyone relax and make sure you are sitting comfortably in a position that will let you fully concentrate for a few minutes . . . If you like, you can close your eyes . . . Now pay attention to your breathing for a moment . . .

> Don't change it—just be conscious of how air is flowing in . . . and out of
> your lungs . . . And now, we can begin with our imagining task.

Once a class has grown accustomed to doing imagining tasks, your introduction can be briefer, until eventually, you may be able to leave it out altogether.

In the next step, you present the exercise to your students. Speak calmly, clearly, and a little more slowly than usual, but at the same time, make sure to keep the suspense going. I highly recommend having the task in front of you in writing so you can read it aloud. Doing so reduces the risk of getting tongue-tied and tripping over your own words. The tone, stress, and pace of your words are key factors in helping students evoke the intended mental images. If you imagine the described content yourself while reading it, this will help you find the right intonation and pace.

Once you've completed the exercise and given sufficient time for students to explore their mental image, you'll close every exercise in Mathematical Imagining with two questions: one about mathematics and one that speaks to the students' individual ways of constructing and experiencing their mental images. The first question varies, depending on the specific mathematics in the task; the second question is always, "What did you imagine during this exercise in Mathematical Imagining?" (We'll explore these questions in detail through examples that follow.) Students have one or two minutes to consider and work through both questions mentally. If possible, they should keep their eyes closed during this step. In it, students are recapping their own, individually constructed math-related mental images and are already beginning to modify or further develop them. When a student is able to accomplish this cognitively challenging feat, the inner stage of their imagination is set to independently and productively use their individual mental images to explore mathematical questions.

The imagining phase ends with you telling your students that it is time to "come back to class" and that they can open their eyes.

The discussion phase: Sharing individual mental images

After the imagining phase, students are eager for verification and want to compare their own mental images with those of their classmates. As a result, students might make spontaneous notes or sketches at the beginning of the second phase and naturally begin talking to one another. To get the hang of this phase, you might invite them to share by saying, "Turn and talk with a partner or your group about what you imagined. Feel free to draw or sketch if that's helpful."

Figure 2.2 shows some student sketches that emerged after they had done the problem-solving exercise PS2, which is about a ladder leaning against a wall. You can find it in Chapter 6, on page 132, and it would be well worth your time to take a few minutes to read it, pause, and imagine it yourself: What did you imagine? If you didn't turn to the exercise, the short version is:

> Imagine a ladder leaning against a wall that's perpendicular to the ground. The ladder has a light on the middle rung. Give the ladder a nudge so that the top slides down the wall. What does the path of the light look like?

If you are using the minimal version of a task, your students' curiosity and need to share are satisfied immediately after the imagining phase through a discussion

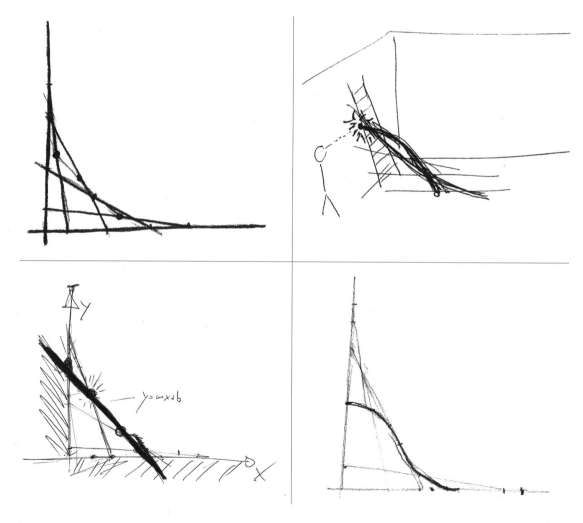

FIGURE 2.2   Student sketches resulting from their individual mental images (the formula in the lower left frame is $y = mx + b$)

round. In it, students describe their experiences and individual mental images to one another in small, self-moderated groups. The mathematical question you posed during the imagining almost always leads to vibrant discussion as well. For example, in the ladder task, you might ask, "What does the path of the light look like when the ladder slides?" Students experiment and argue. They make conjectures and then go on to try and prove and/or disprove one another's speculations. In this process, the discourse seems to innately flow back and forth between the question about how students imagined the task and the suppositions they made about the mathematical question. During this phase, you are more of an observer who is "on the outside looking in." Only rarely will one of the student groups ask you about your own mental images in this phase.

After a while, though, the open, small-group discussions start to wind down, and the entire class will turn toward you. Most of the time, this is when your students expect you—as the expert—to explain the mathematical question. You can choose to meet this expectation, clarify the math question they have been working on, and then pose follow-up questions for more class discussion. (Part 2 of the book contains ideas for follow-up questions to each imagining task.)

This is the point where you decide how far to take an imagining task. Instead of proceeding as described earlier, you can avoid commenting on your student's imaginings or confronting them with the mathematical "textbook" answers—and thus allow the exercise in Mathematical Imagining to reveal its full potential. Your students do not just receive and accept your expert explanation; rather, they delve more deeply into their own imaginings and develop them further. This is where you, as their teacher, now have the chance to learn a great deal about your students' Mathematical Imaginings and the paths they took in the imagining process. This is the expanded version of imagining tasks, which I will now illustrate with a concrete example from practice.

Expanded version: Further developing individual mental images

Individual mental images tend to blur and fade away quickly when exposed to the light of reality. This is a hindrance if you plan to not only discuss your students' imaginings but also further develop them over a series of lessons. A helpful tactic is to ask your students to target and specifically recall what they imagined and then record it in writing. This is why there is an additional phase in the expanded version called *journaling*. It lasts about ten minutes and takes place between the imagining and discussion phases already described in the minimal version (Figure 2.3).

FIGURE 2.3 Phases of exercises in Mathematical Imagining (expanded version)

The following example is from a lesson I did with one of my own classes. I guided my students to view and use the journaling phase as a technique for reflection and knowledge acquisition. This classroom approach was based on the principles of "dialogic learning," as described by the Swiss educators Peter Gallin and Urs Ruf (2005).

The journaling phase: Documenting individual mental images
I used the ladder task (PS2, page 132) to introduce the mathematical concept of "circles and ellipses as trajectories." Right after the imagining phase, I asked my students to document and evaluate everything they had imagined. Here are the instructions I gave them:

a. What is the shape of the trace of light that is "drawn" by the light bulb as the ladder starts sliding away from the wall? Describe your conjectures and draw a sketch.

b. Note down all your mental images—both the pictures and actions—that you imagined during the exercise.

c. Which of your mental images and actions in (b) were useful with regard to your conjectures in (a)? Which ones hindered you?

After my students had finished this part of the task, we had extensive source material available in the form of sketches and journal entries. You have already seen four representative sketches in Figure 2.2. Here, I would like to present and briefly discuss a few complete texts the students wrote in their journals (translated from German to English). These documents reflect student thoughts and processes that I have seen time and time again during my years of working with imagining tasks. The texts from Tanja, Jonas, and Luca (names have been changed) are therefore typical.

the curve of the light bulb

FIGURE 2.4
Tanja's journal entry:
mental images can be
influenced

Ⓐ Die Kurve erinnert *kleinen* an Exponentialfunktionen, geht aber nichts ins Unendliche.

Ⓑ Mein Raum ist im ersten Stock. Quadratisch, weisse (*aus Blech oder so etwas*) Wände, einen grau braunen Teppich. Die Leiter stand zuerst an einer falschen Wand. Ich musste sie holen und an die linke Wand (in der die Türe ist) stellen. Zuerst stellte ich sie ganz nah und ich hatte kein Platz zwischen Leiter und Wand. Dann stellte ich sie weiter weg und mich dazwischen. Als die Leiter dann langsam wegzurutschen begann wollte ich nicht weg-gehen und die Leiter rutschte mir über den Kopf. Ich musste ~~weiter~~ die Leiter anhalten, wegtreten und sie dann weiter rutschen lassen.

Ⓒ hinderlich: – Leiter an falscher Wand
– zuerst war das Licht an
– Kopf zwischen zwei Sprossen
  ↳wegtreten
– nicht unter die Leiter stehen!! ☺

nützlich: – Wände weiss, leer
– Raum quadratisch, leer.
– keine Geräusche oder andere Personen
– Ich konnte Einfluss nehmen auf Umstände, Licht löschen etc.

Ⓐ The curve reminds me in some ways of exponential functions but doesn't continue to infinity.

Ⓑ My room is on the ground floor. Square, white walls, gray-brown carpet. The ladder (made from metal or something like that) was first leaning against the wrong wall. I had to go get it and place it against the wall to the left (where the door is). First, I placed it very close and I didn't have any room between the ladder and the wall. Then I placed it further away and stood in between. As the ladder then slowly began to slide, I didn't want to move away, and the ladder fell around my head. I had to ~~once more~~ stop the ladder, step away, and then let it continue sliding.

Ⓒ Counterproductive:

– ladder on the wrong wall
– at first, the light was on
– head between two rungs → step away
– don't stand under the ladder!! ☺

Useful:

– walls white, empty
– square room, empty
– no sounds or other people
– I was able to influence conditions, turn out the light, etc.

Before discussing the students' work in detail, let's review one key mathematical concept essential to this exercise. The concept captures the shape of a curve that arises, as the trace of the light bulb here, as the graph of a function. The curve is said to be *convex downward* (or *convex*) when the graph is curved like a cup or a hanging rope and *convex upward* (or *concave*) when it is shaped like a hill or an arch. For more details, see footnote 3 in Chapter 6.

Like most students, Tanja spontaneously speculated that the trajectory would be convex downward (Figure 2.4). At first, it reminded her of the graph of an "exponential function," but she quickly revised her statement and mentioned the trajectory remains finite.

What's remarkable about Tanja's journal entry is her concluding sentence: "I was able to influence the conditions, turn out the light, etc." No matter how counterproductive an individually constructed mental image can be (and Tanja's image of the ladder sliding "around my head," is anything other than a comforting or productive one), they can also be restructured or blocked. Mental images are not set in stone; on the contrary, they can be influenced and changed. This is a far-reaching insight when it comes to successfully working with individual mental images!

Jonas also seemed to be "intuitively" imagining a convex-downward trajectory when he sketched the trajectory as a branch of a "hyperbola" (Figure 2.5). His next sketch showing the various stationary points of the light led him not only to speculate about a "circular path" but also to analyze and understand why he had first imagined the curve as convex downward. The question about counterproductive mental images resulted in Jonas wanting to find out the cognitive cause of the irritating image of the convex-downward curve—and he succeeded. "The image of the hyperbola came to me because I was concentrated on the ladder and not on the point. The boundary of the 'ladder's images' is a hyperbola."

In other words, the reason behind students (and others!) predominantly imagining the light's trajectory to be convex downward seems to be that, in spite of the darkening of the room, students combine several mental images of the sliding ladder at different positions into one "group picture." In this way, the sought-for path of the light bulb is "crowded out" by the envelope curve of the "ladder lines" featured in this group picture. As Jonas recognized what was happening, he was able to suppress his hindering image of this convex-downward "hyperbola" and recognize that the path "is actually a circular path."

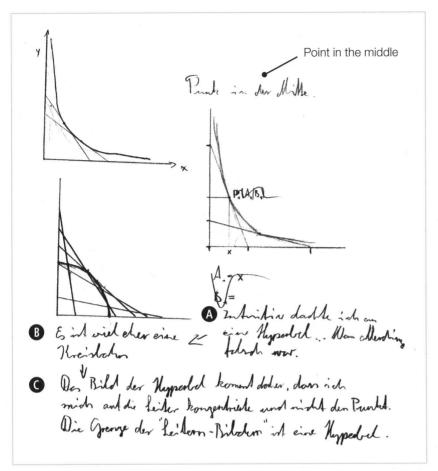

Point in the middle

FIGURE 2.5
Jonas's journal entry:
counterproductive mental
images have their reasons

Ⓐ I intuitively thought of a hyperbola. That was wrong though.

Ⓑ It is actually a circular path.

Ⓒ The image of the hyperbola came to me because I was concentrated on
the ladder and not on the point. The boundary of the "ladder's images" is a
hyperbola.

Luca, like Tanja and Jonas, also imagined "intuitively" that the light's curve would be convex downward (Figure 2.6). After he came up with an image combining the ladder in various positions, though, he started to think differently. He then conjectured that the light's trajectory is shaped like a "quarter circle" from 12:00 to 3:00 (and therefore convex upward). He did not really answer questions (b) and (c); however, he modified the situation in the exercise to let the ladder fall down rather than slide. Accordingly, he formulated the concise and pithy claim that "falling and sliding are the same," which is a productive mental image that goes to the heart of the matter!

With this idea, Luca was able to not only overcome his mental image of the falsely curved trajectory but also succeed in restructuring his counterproductive mental image so that it became a productive one. It subsequently served him—and the entire class—as the viable core concept of the sliding ladder task. With this basis, Luca was able in the next phase to reason argumentatively about why the trajectory is shaped like a circle. Later—and on his own—he proved algebraically that the trace from a light bulb attached to an arbitrary point of a sliding ladder is elliptical in form.

The discussion and reflection phase: Further developing individual mental images and the process of creating knowledge together

Students' journals are an excellent source of initial material for further math lessons. For example, after all students have developed and documented their individual mental images, you can provide them with a forum in which they can express their (justified) interest in one another's mental images and thoughts. This is the reasoning behind the next class assignment: inviting students to take part in a "written conversation."

During this mutual feedback stage, students leave their own journals open on their desks. Then, without talking, they each visit another student's desk located a fixed (not too small) number of seats away from their own and give written feedback on that classmate's notes and sketches. Here were the guiding questions I used in the ladder task:

> Now, have a look at the notes and sketches some of your classmates have made and formulate your feedback:
>
> d. Which mental images do you think are particularly interesting? Why?

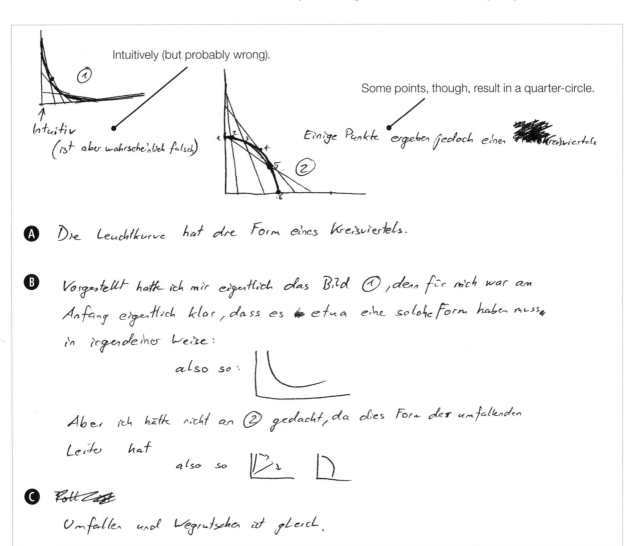

FIGURE 2.6   Luca's journal entry: mental images can be made productive

A The light's curve is in the shape of a quarter circle.

B I had in fact imagined picture 1, because to begin with I was pretty sure that it would have to be ~~in~~ somehow have such a form.

like this:

But I hadn't thought about 2, as this is the shape of the falling ladder:

like this:

C ~~Terrific that~~ Falling and sliding are the same.

e. Which mental images in (d) would help you answer question (a)? In what way?

f. What is your answer now to question (a)? How can you justify your answer?

These instructions are in writing (either on paper to each student or on the chalkboard). Tell your students to give feedback to each other with questions (d) and (e) and afterward answer question (f) in their own journals. When they ponder the first two questions, (d) and (e), students shift from their inward perspective of working alone with their own mental images to an outward, broadening perspective that looks at and considers other people's thoughts and experiences with the task. At this point, the individual mental images of other students form the focus of the student's attention and thereby receive esteem and status. As a result, each student comes to recognize more clearly—and this is crucial—the viability of their own mental images, which is a basic condition for further developing them. After students have considered other perspectives on the task, question (f) then gives them the chance to explore the situation of the sliding ladder again, revising their thinking as needed. Their work with the mathematical question and their own mental images are deepened and take place within a new, expanded perspective.

During the ladder lesson, students worked with these new questions without any mathematical explanations from me. I remained the facilitator and moderator. In order to help fulfill my students' wish for a definitive answer—while still letting the situation speak for itself—I asked two students to demonstrate the exercise on the classroom chalkboard, representing the sliding ladder with a big ruler, and the light bulb with a piece of chalk. This led, at least provisionally, to a satisfactory clarification of the actual path; namely, that the light on the sliding ladder follows a convex-upward curve.

At this point in the lesson, when students have reviewed some of their classmates' work and further developed their own mental images, you can begin the process of creating collaborative knowledge. To achieve this, collect the students' journals after the experiment and select examples to copy or scan and hand out to the class in the next lesson. When choosing student work, you will need to rely heavily on your professional expertise. Keep in mind that you are not primarily looking for entries that exhibit accurate trains of thought or representations that look like textbook case examples. Instead, keep your eyes open for student work that contains a creative core beyond mathematical rules. Include in your selections students' insights and questions and, equally, their (mis-)conceptions or ingenious mistakes. For example, in my class, I included the entries from Jonas and Luca presented in Figures 2.5 and 2.6.

When you distribute the selected student work to your students, give them a new assignment.

> Our experiment on the chalkboard shows that the trajectory of the light bulb attached to the ladder takes the form of a convex-upward curve.
>
> a. Develop other mathematical arguments that support the trace of the light path having this form.
>
> b. Determine the exact mathematical form of the light path.

Students work on this assignment using the basis of the experiment and the selected student documents. Their process will be effortful, as they work to construct meaningful, collaborative knowledge, step-by-step, as opposed to you spoon-feeding them your mathematical knowledge.

In my class, Luca's claim that "falling and sliding are the same" was taken up by many students as they argued for the circular form of the light path. Luca went on to support this argument by comparing the two ladders to a pair of scissors. He went even further and recognized that many fall-front desks (also known as secretary desks) use the same underlying operating principle. In Figure 2.7, the part of the hinge that fastens the fall-front desktop onto the guiding rod correlates to the upper half of the ladder.

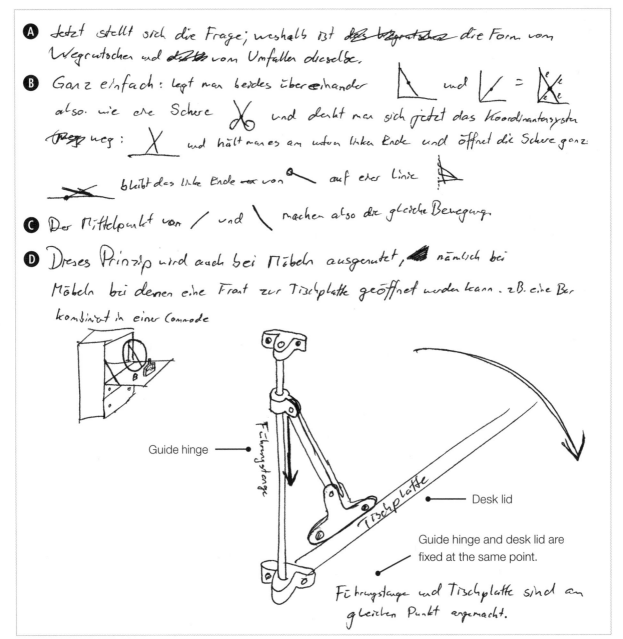

FIGURE 2.7   Excerpt from Luca's journal: Productively restructuring the ladder with a pair of scissors

Ⓐ The question now is: Why are ~~the sliding~~ the shape of sliding and ~~by~~ of falling the same?

Ⓑ Simply if you place them on top of each other ⊾ and ⟋ = ⟋, so like a pair of scissors ✂, and then if you imagine the coordinate system isn't there: ⟋ and if you hold the bottom of the left blade still and open the scissors wide ⟍, the left end of the upper blade ⌒ remains on a straight line ⊿ .

Ⓒ So the midpoints of ⟋ and ⟍ make the same movement.

Ⓓ This principle is also applied with furniture, ◢ namely in furniture pieces that have a front lid that can be opened and dropped down to use as a desktop, for example, like in a secretary desk.

Another one of my students, Tom, was thinking in a similar direction. He used a rod (or as he called it a "pole") to connect the junction of the wall and floor with the middle of the ladder. In his model, the movement now affects the rod and not the ladder. However, Tom still was not satisfied. As he put it, "Why does the room have to be static, while the ladder gets to move?" Subsequently, he carried out a radical restructuring and swapped the roles of the ladder and the room. Tom suggested holding the ladder in place and rotating the room around it, as shown in Figure 2.8. This is an unexpected point of view, and Tom is likely not aware of its productivity. His restructuring work actually comes close to Thales's theorem and thereby to establishing a fixed distance between the light bulb and the corner where the wall meets the floor at right angles. One more role reversal between the ladder and the room would result in the sought-after quarter-circle form of the light path.

At this point, I used these two—and the other remaining student documents—to create a second collection of student manuscripts. Now more students were

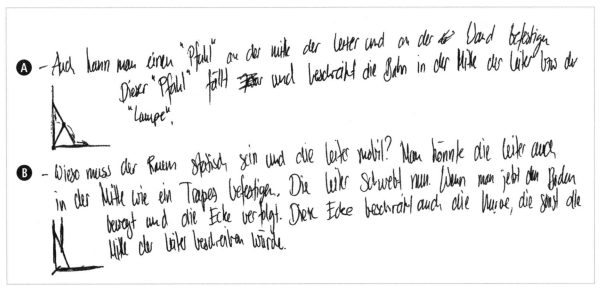

FIGURE 2.8 Excerpt from Tom's journal: productively restructuring by moving the room instead of sliding the ladder

Ⓐ — You can also attach a "pole" to the wall and the middle of the ladder. This "pole" falls and describes the trajectory in the middle of the ladder, that is, of the "light bulb."

Ⓑ — Why does the room have to be static while the ladder gets to move? You could also attach the ladder in the middle like a trapeze. The ladder is hanging then in midair. If you now move the floor and watch the corner: This corner now also depicts the curve, which would otherwise be depicted by the middle of the ladder.

able to make use of their preliminary work, move away from their original convex-downward curve, and prove that the light's trajectory took the form of a convex-upward circular arc. After several false starts, Luca even managed to prove that a light bulb attached to the ladder, but not at its midpoint, moves on an elliptical curve. This is particularly impressive because Luca (and the rest of the class) had no prior experience with elliptical equations.

Although this example visit to my classroom has demonstrated some of what the expanded version of an imagining task can achieve, there is still a great deal of potential contained in students' mental images. So, how can your teaching tap into that potential even further? As a rule, the majority of student contributions are rich in (not necessarily correct) content, so it is worth examining some of their work with a focus on the mathematics. With my students in the lesson exemplified here, I took up my learners' diverse mental images and new ideas and arranged them systematically by topic for future use. For example, in one follow-up task, students used Tom's restructuring of the problem to derive and prove the circular form of the light path.

Perhaps you are growing concerned that the expanded form of the routine will take you outside the content you are supposed to teach and beyond the time within which you are allotted to teach it. And sometimes it may! You will need to constrain Mathematical Imaginings in terms of time, content, and scope and make instructional decisions based on your context and constraints. It might help you justify time spent on Mathematical Imagining, however, to realize that students have to model and formulate and argue and prove, just like in more traditional mathematical exercises, while they learn prescribed content. Evoking and utilizing mental images is one way I teach what I am required to teach in the time I have to teach it.

KEYS FOR SUCCESS

In this section, we'll explore several important factors that are keys for successfully teaching the Mathematical Imagining routine. Please view these as orientation aids and suggestions—and not as hard and fast rules.

Imagining tasks require your mathematical expertise and teaching competencies

The comments in the upcoming exercises will acquaint you with some of the surprising ways of imagining and thinking students come up with. Just like when you use other unstructured, open tasks in your teaching, implementing exercises in Mathematical

Imagining in your classes means lessons that are less predictable and, thus, not as plannable as those that present or require specific procedures and methods.

It goes without saying that you do not have to take up every evoked mental image, especially not those that are merely associative or decorative in nature. The challenge for you is more to ask yourself: *Which mental images contain a productive mathematical nucleus?* This is where you need to bring your own mathematical expertise and teaching competencies into play. You should not merely settle for the commonly accepted "textbook" solutions and paths to reaching them. Instead, you'll draw on your knowledge of various ways a particular learning topic can be approached and understood. This can be quite challenging, but we can help ourselves by allowing sufficient time to explore the imagining tasks ourselves or with colleagues first, noticing our own mental images and anticipating a variety of ways students might imagine.

Exercises in Mathematical Imagining require your curiosity and interest in the origin of knowledge

First and foremost, imagining tasks focus on how solutions are generated, with the solution itself in second place. Therefore, a key to success is being interested in all individual mental images—including your own. When using imagining tasks, you need to take on the role of what I like to call the *curious researcher*. Further, if you are not convinced by the meaning and objective of this teaching tool, then you will not be able to use it effectively. If your learners' mental images bore you, your class is likely to be bored, too. In such cases, the exercises are reduced to a demanding form of brain training with little mathematical value.

Being a curious researcher means, among other things, keeping your own subject-specific knowledge on the back burner until after your students have presented their mental images (while appearing attentive and interested throughout). This approach also includes resisting "improvement mode," in which you act like a pair of scissors, snipping away at anything "extraneous" or "wrong." The tasks are not so much concerned with questions like, *How many sides does a triangle have?* Rather, these types of questions deal with such things as, *How does a triangle look in your mind's eye? How do you imagine it?*

Students find it much easier to accept this type of approach than we teachers do. Particularly when we view ourselves as being primarily specialists of mathematical knowledge, it is quite difficult to deal well with this new teaching tool.

So try to be curious about the wide variety of mental images that come your way. It's a chance to experience your students' creativity and, again and again, see how unexpectedly productive their mental images can be. They can also open new mathematical views to teachers and point us to surprising approaches to math, regardless of whether we are old hands at teaching or just starting out. Freudenthal (1978) emphasized the significance of this attitude when he said, "Moreover I wish to observe learning processes [of children] to improve my understanding of mathematics" (167).

Exercises in Mathematical Imagining require respect

In order for exercises in Mathematical Imagining to work, those participating in the tasks—teachers and students—must take each other seriously and demonstrate mutual respect. Mental images can only be shared when these two factors are a given.

Respect also expresses itself in that the individual mental images are not looked at from a faultfinding point of view or as material for summative assessment. You, as the teacher, should consider and understand your students' mental images from an encouraging and development-oriented perspective, more as material for formative assessment. Here, you should even avoid giving an interpretation. When I asked my students if anything bothered them about the routine, one of them told me, "When there's too much talking about them [the mental images]. Everybody has their own thoughts!" Finally, respect can also mean that, once each school year, you have a class discussion about whether students would like to continue learning math in this way.

Imagining tasks require time and awareness

Constructing, noticing, retaining, and eventually keeping a written record of individual mental images of mathematical subject matter is demanding and requires high levels of awareness and concentration. (When students arrive in your class and are boisterous or exhausted, it is a good idea to begin the imagining task with a brief concentration exercise (see pages 18–19). This not only helps students direct their attention to the imagining task but also helps set the slower pace of the lesson.

During imagining tasks, students work partly on their own with the mathematical question, they experience how the question impacts their own mental images, and they pose their own questions. This process runs counter to a teaching culture that prescribes "just get through the curriculum." Doing imagining tasks under time pressure does not work; the exercises need a certain amount of time. However, my experience has shown that investing fifteen to twenty minutes of a lesson once per week or once every other week in an imagining task is beneficial, as every task can stimulate an intensive and meaningful learning process.

Even when all the conditions necessary for success with imagining tasks are present, challenges and difficulties can still arise. In the last section of this chapter, we'll look at what problems you might face and ways of coping with them.

POTENTIAL CHALLENGES AND SUGGESTED APPROACHES

The teaching tool's cognitively demanding nature means that imagining tasks are occasionally not experienced as challenging but as overwhelming. Some challenges appear when you are establishing the routine, and you can address them then, once and for all. Others though, will keep coming up even after your students have had a longer time to get used to doing imagining tasks. It is important to emphasize, though, that these potential hurdles are absolutely not barriers to students' engagement in Mathematical Imagining. So let us look at some of the potential challenges and suggestions for overcoming them should they occur.

Initial challenges

The first time you introduce a class to imagining tasks one or more of your students might start laughing. Mental images might not seem at first glance to have anything to do with their conception of math, or their ideas could seem too private or even embarrassing. But not to worry, this feeling of insecurity disappears when your students have experienced the cognitive side of their own mental images. They realize that only this cognitive aspect is relevant and going to be addressed. Additionally, they are always in control of which mental images they want to share—or not.

Most likely, for the same reasons, a few students may take a wait-and-see attitude and seem rather reserved during the first imagining task. They observe what the others are doing, what they imagine, and what they disclose. This is not really a problem, as these students will increasingly become accustomed to the new classroom activity and open to imagining tasks. Little by little, all of your students will start engaging with their own mental images, actively participate, and begin to start thinking about mathematics.

Unreasonable expectations are another challenge you could encounter at the start. Students may assume that their mental images have to be razor sharp and film quality as they appear on a kind of inner monitor. The blurriness and vagueness of the mental images that really emerge at first do not encourage students to trust in the power of their own imaginations. Students with high expectations of their own performance especially tend to give up and shut down: "I can't imagine anything!" is a response you might hear in such cases. However, this perceived difficulty with imagining disappears as soon as students gain experience with the teaching tool and learn to trust it.

The hurdles mentioned thus far are minor ones as opposed to the real challenge: being able to imagine the presented situation and questions.

Challenges with imagining and counterproductive mental images

Sometimes students misunderstand the instructions and imagine something entirely different than what was asked for or they will not be able to make their mental images productive. When this happens, they will say things like, "I got stuck" or "I got lost." Most of the time, they will not wait to express their difficulties with imagining until the discussion phase but will signal it immediately with body language during the imagining phase itself. You can observe such occurrences, such as when a student is restless, begins to move around in her seat, or opens his eyes.

As opposed to the initial challenges just described, problems with the actual imagining reappear—regardless of how much trust students have gained in the routine. In Chapter 3, we will look at how to create exercises in Mathematical Imagining that minimize this risk. However, it is not possible to completely avoid the scenario of students having problems with imagining. It is just part of the nature of the beast that students' (or anyone's) own mental images will diverge to a degree from images the instructions intend to evoke. Also, it is not possible to produce mental images that are perfect reflections of the situation described in words. Remember, though, that the effort students make to achieve the desired mental images is one of the positive demands this teaching tool makes, whether or not their resultant image is complete. The following measures can help you support your students through difficulties when imagining:

- Sometimes, you foresee that the mathematical content of an imagining task is going to place high cognitive demands on your class. This can happen, for example, when a task requires preknowledge or experience or language that not all your students have at their disposal. In such cases, you might want to give a brief explanation about the content or portray it graphically with sketches or models. In this way, you alleviate the problem with understanding. However, at the same time, you also take away part of what makes imagining tasks so attractive: getting acquainted with something previously unseen. It is worth considering here if it might not be better to present the mathematical subject matter with a more traditional method. But if you choose to work with such content in an imagining task, I recommend that you plan for a few interruptions. For instance, during the imagining phase, include a moment for a clarifying discussion to ensure that students have understood the instructions. After such a break, guide your students back to the lesson's imagining task.

- No matter how well you plan, though, difficulties with imagining can occur—and often, without you having been able to predict that they would pop up. Imagining difficulties make themselves known through students' obvious non-verbal signals (restlessness, open eyes). You can reach the following agreement with your students so that their body language does not disturb their classmates. When a learner experiences difficulty with imagining, they should quietly raise their hand. This shows you that they are not able to construct mental images and actions for the initiated imagining task but shouldn't distract the rest of the class. You can repeat or paraphrase the part of the instructions you have just given or briefly summarize what you have read so far. In this way, your students who are encountering difficulties with imagining receive another possibility to develop mental images, while those who were managing well can use the time to check the mental images they have already constructed.

- Every now and then, an exercise in Mathematical Imagining overwhelms the entire class. In this case, you can make the written version of the instructions available to your students so learners can approach it at their own pace. You might have them read the text to each other in partners or small groups. When the actual text is in front of everyone, it should be easier for you to locate where the difficulty is and then move on to developing the intended mental images.

Difficulties with imagining do not only appear when students are not able to create mental images of a described mathematical content or only partially able to imagine it. On one hand, imagining tasks do intentionally refer to prior knowledge.

On the other hand, students' individual knowledge base and life experiences can be at the root of their imagining problem and act like a mental roadblock. There are differing reasons behind the appearance of such *counterproductive mental images*:

- When the nonmathematical context prompts memories that entice students to wander off and leads them away from the intended mental images. This is likely the most common reason behind student difficulty with imagining.

- Counterproductive mental images show up regularly when students manipulate their mental images. When they zoom in on certain parts of a mental image, the image as a whole fades to the background. When they then try to zoom back out from a particular part and reconstruct the original image, it suddenly looks entirely different from the one they had faded out. For example, a cube can mutate to a pyramid when a learner starts focusing on the three edges joining at a corner of the cube.

- Just like in visual representations, we do not perceive all aspects of mental images at the same time or assign them the same weight. Some elements push their way into the foreground, such as the ladder lines in the prior example, or—although we know better—we "see" some characteristics as belonging together.

- When mental images are eidetic, meaning they appear too realistic and are experienced like real visual perceptions, students can have increased difficulty in working with the images mentally. The specific images are so realistic that students are hardly able to manipulate them and answer the mathematical question.

- Finally, individual mental images can also bring up emotional, negative associations that cause the task to fail. It is no longer possible, then, for students in this situation to autonomously deal with their own mental images, and the imagining task must be stopped. As an example, should students imagine themselves standing over a dark, seemingly bottomless pit they could become quite anxious.

All of these counterproductive mental images are not the same as misconceptions, in which the imagined content is not correct from a mathematical point of view. In contrast, counterproductive mental images can be inappropriate from a mental point of view, yet still contain aspects that can be made mathematically productive. When you judge students' mental images according to whether the math within them is "right or wrong," you are using a deficit-oriented perspective. The approach you need to take is one that is development-oriented, meaning you view students' mental images according to how productive or counterproductive they are. Accordingly, counterproductive mental images do not need to be deleted or discarded. Instead, students can work with the mental images that are hindering them and, through manipulation or restructuring, make them productive, like Luca (Figures 2.6 and 2.7) and Tom (Figure 2.8) did. Students can learn that they do not have to accept a counterproductive mental image as is, and neither do such mental images have to stay the way they first appeared forever. Indeed, this transformation is empowering.

Should a counterproductive mental image dominate to such an extent that it resists transformation, students can construct another mental image and partner it to their first one. For example, this is what Luca did when he complemented the sliding ladder with the falling one and also by partnering the image of the ladder with the one of the scissors. In this way, the counterproductive mental image will at least lose its dominance. This kind of learning can quite often guide learners—and their teachers—to new insights during their adventure with an imagining task.

CHAPTER

3

Developing
Your Own Exercises
in Mathematical Imagining

Creating an imagining task requires you to express specific mathematical content in vivid, clear, and easy-to-understand language. This does not mean putting a formula or proof into words, as imagining tasks are not lectures read out loud from the textbook. In this chapter, I'll teach you how to develop your own Mathematical Imagining tasks, and we'll consider what subject matter is particularly suited for use with them.

There are three aspects you need to consider when creating your own exercises in Mathematical Imagining: the *mathematical content*, the *language usage*, and the *sequence of text elements*. We'll take each aspect in turn, generating a list of the traits common to effective imagining tasks. Before we continue, though, please note two important considerations:

- This discussion of what makes for a quality imagining task should not inhibit or deter you from creating your own imagining tasks. It is not possible for every task to fully meet every ideal, and the various attributes are not equally important for each task. Depending on the exercise, trying to incorporate all the traits can cause unsolvable conflicts and even be contradictory. The point is not so much to have tasks that optimally comply with each trait but, rather, ones that fulfill the overriding vision as best as possible.

- Do, though, follow the suggestions when creating your first imagining tasks. As soon as you have gained a bit of experience, you will know what the key factors are and will want some flexibility. As you continue to create Mathematical Imaginings, you will likely go on to identify your own additional criteria for what makes a quality task!

How do exercises in Mathematical Imagining originate and develop? What, for example, is my own approach? Well, ideas for imagining tasks only rarely come to me while I am working at my desk or when I am consulting textbooks, as these stimuli easily get in the way of my own imagination. Most often, an idea for an imagining task emerges when I am mulling over a mathematical subject while doing something else, like relaxing on the sofa or riding my bike. I spend some time clarifying the idea to see whether it's mathematically productive. If it is, and once I have made the mathematical idea I want my students to work with clear to myself, I then look for a suitable context I can use to visualize it with and draft the first version of the task instructions. Next, I share the draft with a friend, preferably over the phone, and afterward I rework and fine-tune the linguistic formulations. Only then is the exercise in Mathematical Imagining ready to use with my students.

THE MATHEMATICAL CONTENT

Experience in my classroom and leading professional development has shown me that imagining tasks work especially well when they arise and develop from the teacher's own imagination, contain simple, yet rich content, take the students into account, and are formulated in the most natural and unaffected language possible. Let's begin with content.

Simple, yet rich content

Imagining tasks take up *simple* and at the same time *rich mathematical* content to reach their overriding aim, namely, appeal to listeners with math-context questions that then initiate mathematical processes (Table 3.1). "Simple" here in no way means

TABLE 3.1   Characteristics of mathematical content in effective imagining tasks

Simple, yet rich content	• Mathematical idea is inviting: phenomenon or thought-provoking problem • Appropriate cognitive complexity • Not a mere rehash of familiar knowledge • No pedantic discussions about mathematical terms
Content can be visualized	• Content can be geometrically implemented • Sizes only as magnitudes (ballpark figures) or proportions • No exact or absolute specification of sizes
Content can be embedded in an everyday context of objects and actions	• "Real" objects and actions • Context is kept throughout the imagining task • Students observe and act within the context • Students can "be in" the images and mentally perform actions as themselves on their individual inner stage • No instructions to manipulate mathematical symbols or rearrange formulas

working with banal content or rehashing students' old knowledge that they are already well acquainted with. The point is more to choose content with a level of cognitive complexity that is neither too low nor too high. The content of an exercise in Mathematical Imagining must be easily accessible and challenging at the same time.

This requirement is part of the demanding nature of the routine itself. After all, Mathematical Imagining calls for students to imagine specific content without any help from sketches or models. The less attention students have to pay to minutiae while they are imagining, the easier it will be for them to get the mathematical idea. Therefore, effective imagining tasks are built on simple *mathematical phenomena* and on *thought-provoking problems*. Tasks with such subject matter induce thinking and reflection, trigger engagement and interest, and facilitate students having their own personal access to mathematics. When it comes to the content, less is more.

Content can be visualized

The next condition that applies to the mathematical content is: Can it be *visualized*, and if yes, to what degree? The very nature of geometric constructions, problems, and reasoning means they all meet the "visualization" requirement. Content from topology is particularly well suited for imagining tasks then: mental images are not exact enough to be measured or calculated, and topology deals with geometric properties that are beyond measuring and calculating.

The mathematical content does not, though, necessarily have to come from geometry. It suffices when the content can be geometrically implemented and visualized. For this reason, an imagining task can take up content from arithmetic (e.g., number patterns, problem-solving task PS5), algebra (special binomial products, reasoning task R5), or calculus (convergence of a geometrical series, reasoning task R2). However, I must emphasize strongly that the requirement that content must be visualizable has a consequence: you will not be able to make all subject matter in your curriculum a topic for an imagining task. There is no need to try; Mathematical Imagining is one teaching tool among many.

The visualization factor is particularly important when you wish to use content that typically involves exact or absolute size indications. Imagining such instructions is problematic, so you will need to use qualitative arguments to overcome this issue. For example, rather than saying a segment is 4 inches, you might say the segment is "half the length of my hand." In other words, rather than giving an absolute measurement, you can give a relative measurement (such as in reasoning task R3). As students learn to scale or squeeze their mental images of situations at will, it is generally sufficient to use expressions like "the width of your hand" to describe a measurement or "a third of the rod" for proportions.

Content can be embedded in an everyday context

The third, essential characteristic of the mathematical content of imagining tasks is that the subject matter must lend itself to being imagined with *real, everyday objects and actions*. Further, the descriptions are not limited to the objects, as the instructions also transport the students personally into the context. Here, students—in a sense—interact with the content as themselves and participate as actors in their mental images. Equally important is that every exercise in Mathematical Imagining also guides students in performing *mental actions*. As discussed in Chapter 1, these mental actions are inherent to the dual nature of imagining itself. They can even go so far that students feel called to go beyond working with and manipulating their mental images and actually move themselves within the realistic context. This brings dynamism into math class!

This embedding of the mathematics in an everyday context should not seem artificial or far-fetched and also should not suggest that math is being used to solve a real-life problem. Instead, the everyday context should make the mathematical question more accessible, so the threshold for entering the imagining task should not be set too high. The everyday setting also neutralizes many biases against math because, at first, students are working in a nonmathematical context. Then they can successively explore the math content, formulate their assumptions, and ask their own questions. However, they will not be able to enter a task if your instructions ask them only to manipulate mathematical expressions, for example, "Imagine multiplying by x." Such instructions have a completely different aim than an imagining task.

As we've seen, successful Mathematical Imagining tasks involve (1) simple, rich content that (2) can be visualized (3) in an everyday context. How strictly you need to adhere to these three descriptors depends on your students' readiness. An exercise in Mathematical Imagining must be tailored to each class, and you should not overestimate what students of a particular age are able to imagine. If you base an exercise on prior knowledge and experience that your students do not possess, then difficulties with constructing the intended mental images are bound to happen. Therefore, I suggest starting and becoming fluent with tasks that fulfill all three criteria before experimenting.

LANGUAGE USAGE

A university student of mine put his finger exactly on the point about the crucial role word choice plays in imagining tasks.

> *A mathematician tends to make steps that are too big. When words should be transformed to mental images, then each and every word has to be understandable so that every statement can be interpreted immediately. The listener has to be pulled along a thread, one step at a time. Otherwise, they will fall by the wayside.*

Because students receive the content of an imagining task through the means of the spoken word, the language you use plays at least as decisive a role as the choice of content matter. Even when an exercise in Mathematical Imagining fulfills all the content requirements, if you do not phrase the task appropriately, students will not be able to imagine it. (Table 3.2 provides an overview of the characteristics of language usage.)

TABLE 3.2   Characteristics of language usage in effective imagining tasks

Descriptive and concrete language	• Everyday, informal language • Recalls real-life experiences and general knowledge • Language close to students • Addresses the individual student • No formal or expert language • No trivial or chummy language
Analogies and metaphors	• Metaphors targeting visual, tactile, motor, or movable mental images • Figurative and concrete words • Not just provide "given" descriptions, also promote students "doing" • No unnecessary details or overexact descriptions
Main verbs instruct mental actions	• Verbs target the action to be imagined: mental manipulations and dynamic mental movements • Direct and concrete instructions with imperative verbs • No impersonal or passive formulations, no use of conditionals

Descriptive and concrete language

The first responsibility of every imagining task is to present a situation that can be envisioned. The text should not sound like code students have to decipher. Exercises in Mathematical Imagining should be spoken in clear, everyday, and concrete language, without abstract terms. This language refrains from using mathematical terminology that—when viewed with a bit of distance—our students are most likely not familiar with. Instead, use age-appropriate language that is close to the way students talk. This does not imply "chumming up" to your students or trying to use their slang. Rather, in order to situate the mathematical content in a non-math context, the imagining task draws on students' everyday experience and prior knowledge. For example, you can recall students' prior knowledge, such as "two points in the plane can be connected with a line segment" but not knowledge of unfamiliar theorems, like "due to the intercept theorem . . ." Purely logical arguments and wording such as "contradictory to" are also not fruitful within the framework of an imagining task.

In addition, when choosing your wording, remember that during an exercise in Mathematical Imagining, each student is working independently. So, make sure to address the class as individuals and not as an entity. This is particularly important in how you actually read instructions, such as asking, "What did *you* imagine?" Your intonation and emphasis need to make clear that such questions (or other remarks) are individual—and not collective ones.

Analogies and metaphors

Linguistically, it's important you embed the math content within a context of objects and actions. In particular, you need to make optimal use of the *metaphorical quality* of your words.

- Metaphors are words, idioms, and expressions with figurative meaning. When creating and learning mathematics, metaphors play the role of mental bridges and help establish analogies. Metaphors can compress into one single picture multiple facts that could otherwise become unmanageable. Skillful use of metaphors and analogies makes the task language more efficient, because students have fewer words to grasp and remember. This compactness is of inestimable worth when it comes to evoking mental images. Metaphors like "in the form of the letter L" or "a triangle with wings" suggest pictorial or visual mental images. You can also use other metaphors in imagining tasks that target tactile or motor images, for example, "a massive rock" or "a hinge." In principle, you can even target mental images that draw on other senses (auditory, and so on); however, this happens only in rare cases.

- Although using metaphorical language supports the construction of mental images, there is a risk involved in using it. There is always only a portion of the pictorial connotation that is desired or helpful. For example, the word *bowl* can help students construct half of an icosahedron in construction exercise C1, but it can also bring up mental images that go in another direction. The image of the bowl contains the idea that five equilateral triangles connected around a point create an object that is not flat but, rather, concave. The purpose of a bowl as an object to contain something, however, is an image that must be ignored during this task.

The conflict referred to is an inherent part of language, and it cannot be bridged by using overexact, detailed descriptions. We teachers, as a matter of course, are always coping with the multiple meanings words can and do have for our students and work with the corresponding misunderstandings. This complexity applies to students' explorations with imagining tasks even more so than when they are learning mathematics in general.

Main verbs instruct mental actions

If the text of an exercise in Mathematical Imagining just evokes one static picture after another, it is one sided and monotonous. We can improve the exercise by making it dynamic, by adding mental actions to the mental pictures. Texts that use *verbs* that direct students to imagine actions situate the mathematical content in a context of actions. These mental actions can be differentiated into two types: *discrete, static movements* or *fluid, dynamic movements*.

- Many geometric figures and objects can be put together piece by piece. In doing this, there is a "before" and "after," meaning there are single, discrete mental pictures that follow one upon the other. In an imagining task, for example, students can construct a cube out of individual squares, one face at a time. Your instructions prompt students to perform a series of mental actions to achieve this, which results in a series of incremental, static images, as in a series of still photographs or snapshots. For instance, their first image shows a square that is lying on their desk. You then ask them to put another square on it and describe the orientation such that their next picture shows two square surfaces that are perpendicular to each other, and so on. In addition to assembling, other examples of mental manipulations that are employed in many imagining tasks include dismantling, cutting off, painting, and bisecting.

- Alternatively, your instructions can lead students to vary an aspect of a geometrical figure, which results in a morphing process on the figure as a whole. These mental movements are normally accompanied by images that flow from one to another, just like in an animated film where one image changes (or morphs) into another through a seamless transition. For instance, the endpoints of a line segment of constant length can be moved along two right-angled axes in order to discover how the midpoint of the line segment moves on a specific curve. (This is the sliding ladder, problem-solving exercise PS2.) Additionally, mental movements like shearing, walking, zooming, and changing perspective all have a place in many imagining tasks. With such features, Mathematical Imaginings allow us to incorporate the dynamic side of math that is often neglected in static textbooks.

In order to bring these movements to life, you'll want to address your students directly and use the imperative for all your instructions about actions, that is, "(you) cut ... off," or "(you) move the ...." You should avoid putting passive or impersonal formulations into your texts ("is to be turned ...," "we will add now ...," "one can see ...," etc.). Such structures are not nearly as inviting as active language that speaks and appeals directly to your students (and their imaginations). You should also stay away from using the conditionals (*if* sentences) to direct students' actions. Giving, "If you ... then ... will/would ..." does not really make sense in the world of imagining task, which is already hypothetical in and of itself.

THE SEQUENCE OF THE TEXT ELEMENTS IN THE INSTRUCTIONS

As soon as a sentence is spoken, it drifts away and begins to disappear. Once you have read the imagining task's text to your students, they cannot rewind and listen again to information they might have missed. This ephemeral nature is why it is important to sequence text elements in your instructions with care. After experimenting with different arrangements, I have developed some criteria that are crucial for the success of an imagining task. (Table 3.3 gives you an overview of these characteristics.)

TABLE 3.3 Characteristics of the sequence of text elements in effective imagining tasks

Chronological and causatively logical order of text elements	• Step-by-step, like a recipe or instructions for arts and crafts • Suspense builds and leads to the mathematical idea • No mental leaps
Short breaks	• At appropriate places, sum up the instructions thus far • Language may include redundancies
Two concluding questions	• First question about the specific *mathematical content*: targets the mathematical idea • Second question about the *individually constructed mental images:* invitation to bring math-related images to the fore

Chronological and causatively logical order of text elements

Put simply, every exercise in Mathematical Imagining follows the same schema of "construction–action–questions." The order of the text elements is closely tied to the chronological and causatively logical sequence. Therefore, you do not mention objects in advance but supply them on demand, meaning at exactly the moment when they become the topic. If you say the name of an object and then wait awhile before describing it further, students who do not know the object's name will experience uncertainty. This is the same reason why you need to avoid making mental leaps or anticipatory comments.

The order of the text elements most closely resembles a recipe or instructions for arts and crafts. You lead students to construct their mental images step-by-step. Ideally, this buildup creates suspense that targets the mathematical idea and culminates in the mathematical question at the end of every imagining task, which we'll discuss next.

Insertion of short breaks

The language of mathematics is spare and efficient. In a Mathematical Imagining task, however, redundancy is important. Include in your instructions a short "making sense" break after each conceptual step so students can catch up before you move to the next. In it, recapitulate what has been accomplished thus far and sum up the present state of affairs. Such repetition helps build students' confidence. When writing your instructions, remember: imagining tasks are not the place to try and adhere to the pure principles of mathematics and liberate them from all extra, nonrelated thoughts or ambiguous meanings but, on the contrary, to use words, idioms, and expressions with figurative meaning, that is, metaphors and analogies.

Two concluding questions

Every instruction text closes with two questions, one mathematical and one that inquires about the individually constructed mental images.

- The *first question* targets the mathematical nucleus contained in the imagining task. It concerns the phenomenon at the heart of the exercise. Usually, the way the task instructions build up will indicate the question, and to an extent, students might even see it coming. The question's formulation is always rooted in the intended mental images, meaning it can be answered by using certain mental images. This question can be so far reaching that it can run like a thread through several following lessons (e.g, the sliding ladder task in Chapter 2).

- The *second question* targets the individual student's mental images and actions. Every imagining task has the same concluding question: *What did you imagine during this exercise in Mathematical Imagining?* Even though this question is formulated in a personal way, the purpose is not to have a tell-all session where students share every single idea or association they had while imagining. This question is more about bringing to the fore the math-related mental images.

All students, at all levels, can answer the concluding question, which allows differentiated instruction and thus leads to math for all because all students can answer it—no matter their achievement or performance level in math. Every single one of your students imagines something and constructs individual, math-related mental images, which at first are neither right nor wrong. There is a broad spectrum of answers, which you will find almost impossible to predict in advance. This is truly a *genuine* and *open* question.

In other words, the question about the individual mental images initiates processes that, at first glance, may appear to be less obviously relevant to your math instruction. However, if you do not ask this concluding question, you will miss out on a substantial part of the imagining tasks' potential (see Chapter 1).

Now that we have some clearer ideas about the attributes of successful tasks, we can turn to example exercises in Mathematical Imagining.

_ P A R T 2 _

Exercises in Mathematical Imagining: A Collection of Example Tasks

An expert problem solver must be endowed with two incompatible qualities:

a restless imagination and a patient pertinacity.

—Howard W. Eves, *In Mathematical Circles: A Selection of Mathematical Stories and Anecdotes*

C H A P T E R

4

About Using the Example Tasks

This part of the book provides you with many example imagining tasks. These exercises in Mathematical Imagining stem from a variety of math disciplines taught at the high school level. Here are two suggestions to keep in mind when working with the provided instructions for the example tasks:

- The order in which you see the imagining tasks here does not imply in any way that you should work with them in this order or even implement all of them in your classes. Browse through the tasks and choose one suited to your classes! The mathematical idea and the prerequisites provided at the beginning of each exercise can help to inform your choice. Refer to Chapter 2 for how to implement exercises in Mathematical Imagining in your lessons, and you will find background information in Chapter 1.

- All the imagining tasks have been used many times with my students in class. Depending on your own intentions and the level of your students' prior knowledge, you may need to customize tasks in terms of content and/

or wording. In such a case, or if you want to create your own imagining task, it is a good idea to have a look at Chapter 3 first.

In the first section of this chapter, I'll describe the four different types of tasks. In the second section, I'll discuss the information and comments that accompany the imagining tasks.

TYPES OF IMAGINING TASKS

At minimum, every exercise in Mathematical Imagining seeks to help students engage with and visualize a mathematical topic. To achieve these goals, the task is framed in a context of objects and actions.

In addition, Mathematical Imagining tasks are effective tools for teaching four types of mathematical activities. Construction tasks make content accessible and ripe for exploration. Problem-solving tasks encourage experimentation and guessing. Reasoning tasks are rich contexts for developing plausible mathematical arguments. Finally, paradox tasks give students opportunities to experience and examine cognitive conflict and scrutinize their own assumptions. For more details, see Table 4.1.

Naturally, there is crossover among the different types. For example, your students can use a mentally constructed mathematical object for reasoning about a mathematical concept, or through a few modifications, you can make the same object the centerpiece of an open task. Therefore, there is no need to be overly rigid about these classifications. My main hope is that categorizing them corresponding to the student thinking they generate will help you select tasks purposefully.

I have organized the tasks based on their types and their dominant mathematical ideas, giving each task a letter and a number code. For example, C7 stands for the seventh construction exercise in Mathematical Imagining. Moreover, each chapter of this part of the book begins with a brief overview.

Because the sequence of courses varies so much nationally and internationally, I have refrained from providing suggested grade levels or course names for the tasks. I do provide substantial information about the mathematical ideas in each task, so you will be able to determine whether a particular task is suitable for your students and your course goals.

TABLE 4.1   Four different types of exercises in Mathematical Imagining

Construction exercises in Mathematical Imagining (C)

Construction imagining tasks give contextual, step-by-step instructions on how to mentally construct a mathematical object or principle. They then prompt students to explore some of its features and answer a mathematical question. These tasks are primarily concerned with *making content accessible and exploring mathematical objects and principles*.

Problem-solving exercises in Mathematical Imagining (PS)

Problem-solving imagining tasks are designed to foster students formulating and conjecturing about their own hypotheses. Examples include the rolling plastic cup or sliding ladder tasks discussed in Part 1. These exercises in Mathematical Imagining do not provide students with a strategy for answering the mathematical question at hand. This is why, in contrast to construction tasks, problem-solving tasks are open. In short, problem-solving exercises in Mathematical Imagining put *experimenting and guessing* at center stage.

Reasoning exercises in Mathematical Imagining (R)

Reasoning exercises in Mathematical Imagining (such as the triangle task from the start of the book) engage students in heuristic processes, including the creation of their own informal, plausible arguments. Even when learners are already familiar with the mathematical idea behind the argument, such as the sum of the angles of a plane triangle is 180°, the task presents it in a new, visualized context, such as walking and rotating around the triangle. These imagining tasks are not axiomatic-deductive, formal proofs. Instead, they present preformal arguments and reveal characteristics and background that add meaning and credibility to the assertion. Reasoning exercises in Mathematical Imagining often trigger an "aha!" effect, and it is exactly this subjective aspect of plausibility that plays a significant role when it comes to accepting mathematical proofs. This is why plausible reasoning should not be neglected in school math classes, and also why, for this type of imagining tasks, the term *reasoning* is preferred to *proof*. In short, exercises in Mathematical Imagining of the reasoning type initiate *plausible reasoning* and complement formal arguments and proofs.

Paradox exercises in Mathematical Imagining (P)

Paradox imagining tasks describe mathematical situations that are not accessible through everyday life experiences and are therefore counterintuitive. They seem to go against the grain of our common sense—and so, are paradoxes. The imagining tasks here trigger what seem to be contradictions (but when viewed logically, are not). In doing so, these paradox imagining tasks are designed for a much different goal than reasoning tasks. Paradox exercises in Mathematical Imagining raise a cognitive conflict that leads to questions or promotes discussions that go beyond students' everyday life experiences and school math knowledge. Mathematical objects, concepts, and procedures are subject to *questioning and reflection* so that students can better integrate and more deeply understand them.

COMMENTS ON THE IMAGINING TASKS

If you look through some of the Mathematical Imagining tasks in Chapters 5–8, you'll see that I comment extensively on each of the example tasks. The comments give you pedagogical and mathematical background, discuss possible difficulties students might encounter, and outline potential follow-up questions. The tasks and comments all have the following structure.

Overview

At the beginning of each exercise, there is information that tells you the difficulty level, the mathematical idea, and any necessary prerequisites.

DIFFICULTY LEVEL:	The relative difficulty level is symbolized in the following way:

★   easy

★ ★   challenging

★ ★ ★   very challenging

The *difficulty level* I have assigned to the imagining tasks is intended to be a rough guideline, only for the exercise itself, and it does not apply to the follow-up questions. This estimate is also relative, as so many other factors come into play here. Students' prior learning, language competence, concentration ability, and the kind of day they are having are just a few of the many dynamics that determine how easy or hard a task will be for them. Naturally, when you use an exercise in Mathematical Imagining to introduce totally new content, then students will find it more challenging than when you use it retrospectively.

MATHEMATICAL IDEA:	What content does this task address? This is where you will find the core mathematical idea.
PREREQUISITES:	A list informs you about skills and prior knowledge that students need in order to access the task:

- Everyday experiences or skills, such as walking along the edges of a triangle, cutting a square piece of paper, constructing an arc with a drawing compass, and so forth.

- Mathematical prior knowledge, for example, thinking of an angle as an invitation to rotate a certain number of degrees, seeing the image of a square standing on one of its corners, or knowing that every plane cross section of a sphere is a circle, and so forth.

Mathematical and pedagogical background

After the overview, you will find a brief section with mathematical and background information about the imagining task. In this section, I describe how the exercise leads to the mathematical idea.

Task instructions

The text of every exercise is offset.

- The instructions are divided into text passages where related task steps and concepts are grouped together in meaningful segments. These sense units appear with a bullet point (•).

- The text designates places where you need to pause for a moment with an ellipsis (. . .).

- Words that you should emphasize are in *italic*.

These formatting elements are there to support you when you read the imagining task instructions to your students.

All exercises in Mathematical Imagining have been implemented in classrooms numerous times, so you should be able to use the instructions with your students as they appear here. Depending on what you want to accomplish with them and/ or the prior knowledge of your class, you may need to modify the wording to suit your situation.

Comments

Right after each actual task, you will find comments about both the mathematics and pedagogy of the task. (The first two imagining tasks of each type have much more expansive comments about the content of the mental images.) This is where I describe the mental images the task intended for students to construct (*Notes on intended mental images*). You will also find examples of the mental images students actually construct in practice (*Notes on the productive and counterproductive mental images that learners construct*). These mental images are not assessed using mathematical criteria (right or wrong?) but, rather, are assessed on whether they support answering the mathematical question, and if so, how.

- *Productive mental images* can lay the foundation for additional tasks that you had already "planned" to have students construct later. They might also have an exemplary nature that can guide further learning, such as when a mental image contains a key to answering the mathematical question. Such creative efforts from your students can serve to help other learners overcome their difficulties with imagining and, thus, contribute to all your students learning through the task.

- Many images that students construct during an exercise in Mathematical Imagining hinder the continuation of the imagining process. Because this teaching tool focuses on evoking your students' individual mental images, it is inevitable that some *counterproductive mental images* will come to light. (For dealing with difficulties with imagining, particularly with counterproductive mental images, see pages 36–39 in Chapter 2.)

This section will help you gain insights into the way learners imagine and think. Of course, it shows only a fraction of what actually goes on in our students' heads. So look forward to discovering what imagining tasks can reveal about how *your* students imagine and think when engaged in math!

Mathematical Follow-Up Questions

The last section of the comments provides mathematical follow-up questions and activities that can be developed from the task. The goal here—and I must emphasize this strongly—is not for you to work through all the questions and points raised with a class or to give them a presentation about the material. My aim here in providing this collection of impulses is more to indicate in which directions the individual mental images can be further developed. This also explains why I neither use formal language to present the information nor include all the answers and proofs.

My final remark here is about the book's illustrations: when I have provided graphics for an imagining task, I have placed them in the follow-up questions section only. I have almost completely refrained from using illustrations in the comments—and never in the actual imagining task texts. This is because I do not mean to imply in any way that your students' mental images should match the illustrations. This is definitely not the point when it comes to constructing individual mental images. On the contrary, as explained earlier, exercises in Mathematical Imagining are designed to nurture independent thinking. They are opportunities for our students to not only experience mathematics as something personal— something that originates and happens within each of us—but also discover that their personal "bits" (images, etc.) are the basic elements that enable students to follow their own paths to mathematical solutions.

CHAPTER

5

Construction Exercises in Mathematical Imagining

In this chapter, we'll explore the construction type of exercises in Mathematical Imagining. These imagining tasks guide students through the construction of a mathematical scenario so that learners are then able to answer a mathematical question. Unlike, for instance, problem-solving-type exercises in Mathematical Imagining (Chapter 6), the tasks here provide step-by-step instructions to support students in constructing their mental images.

Table 5.1 gives you a general overview of all thirteen example tasks and specifies their educational and methodological backgrounds. Each imagining task raises a number of mathematical questions that can also be used to guide the rest of the lesson or future ones. In the first two exercises (C1 and C2), I discuss in detail the mental images these tasks are intended to evoke and ones that have arisen in my own practice. The remaining imagining tasks follow these first two and are sorted according to level of difficulty (more stars denote more difficulty).

TABLE 5.1 List of construction-type exercises in Mathematical Imagining

Exercise in Mathematical Imagining	Mathematical idea	Prerequisites	Pages
Constructing an icosahedron ★ ★	Icosahedra can be created by joining together two congruent objects that each have five attached flaps.	• Regular hexagons consist of six regular triangles (see R3). • Score a piece of cardboard to crease it.	C1 p. 62
Cross-sectioning a cube ★ ★ ★	Cross sections of a cube taken perpendicular to a space diagonal are regular triangles or hexagons.	• Position a cube with one space diagonal perpendicular to the desk surface. • Distinguish between a solid cube and a wire-frame cube.	C2 p. 69
Laying out a harmonic Pythagorean triangle ★	Laying out a triangular dot pattern with ten checkers pieces.	• Place checkers game pieces in a regular pattern.	C3 p. 75
Truncating a triangle ★	The process of cutting off the corners of a triangle ever more deeply results in a series of hexagons and, finally, a triangle.	• Familiarity with the shape of an equilateral triangle. • A trapezoid is obtained by cutting the corner of a triangle off with a line parallel to its opposite side.	C4 p. 79
Constructing a pentagram ★	A pentagram can be constructed by following combinatorial instructions.	• Properties of a circle. • Jumping and skipping while playing board games (e.g., Chinese checkers).	C5 p. 83
Folding a tetrahedron ★ ★	Tetrahedra can be folded from a rectangular sheet of paper without making cuts or overlap.	• Rolling a cylinder tube from a sheet of paper.	C6 p. 88
Constructing a figure of constant width ★ ★	The circle is not the only convex shape of constant width.	• Construct an equilateral triangle with a compass. • Construct parallel line segments.	C7 p. 92
Dualizing a regular tiling of the plane ★ ★	The regular tiling with squares is invariant under dualization.	• Regular tiling with squares. • Center of a square.	C8 p. 96

(continued)

TABLE 5.1 List of construction-type exercises in Mathematical Imagining *(continued)*

Exercise in Mathematical Imagining	Mathematical idea	Prerequisites	Pages
Ambiguous pyramid ★ ★	The ground plan of the wire-frame model of a square pyramid can be seen in different ways.	• Ground plan (or view from above) of a square pyramid.	C9 p. 99
Constructing a dodecahedron ★ ★	A dodecahedron can be constructed out of two congruent objects each with five attached flaps.	• Familiarity with the shape of a regular pentagon. • When five congruent pentagons are glued to the sides of a sixth identical pentagon, wedge-shaped gaps appear between the attached pentagons.	C10 p. 103
Constructing a solid of constant width ★ ★	The sphere is not the only convex solid with constant width.	• Constructing a Reuleaux triangle (see C7). • Rotating a regular triangle in space.	C11 p. 108
Constructing the Cantor set ★ ★ ★	Certain geometric objects are intermediate between a point and a line segment.	• Dividing a line segment and removing a subsegment. • Continuously repeating an action.	C12 p. 112
Constructing a four-dimensional hypercube ★ ★ ★	If you find just the right translation you can build a four-dimensional hypercube.	• Line, square, and cube. • If a luminous point is moved straight ahead, the trace of the light forms a line segment.	C13 p. 116

C1 CONSTRUCTING AN ICOSAHEDRON

DIFFICULTY LEVEL: ★ ★

MATHEMATICAL IDEA: Icosahedra can be created by joining together two congruent objects; each has five attached flaps.

PREREQUISITES:
- Regular hexagons consist of six regular triangles (see R3).
- Score a piece of cardboard to crease it.

Few students have a mental image of an icosahedron because, unlike cubes, which play a starring role in high school math lessons, icosahedra are not as common in core curricula. Learners usually explore icosahedra using a real model (out of wood or wire) or a virtual model on a computer display.

You could introduce icosahedra by describing their properties, for example, as having twenty congruent equilateral triangles joined together, with five triangles meeting at each vertex. This description provides a technical idea of an icosahedron's shape but one focusing on the neighborhoods of its vertices. In contrast, the wording of this imagining task aims to guide students' mental construction processes and make the concept of the icosahedron as a whole available to them.

The instructions to this exercise in mental imagining are formulated like a how-to guide for making a model of an icosahedron out of cardboard. It begins with the mental image of a regular hexagon. Then students mentally use various tools to work on this shape in the plane. They then turn it into a three-dimensional dome shape with a zigzag border, replicate it, and join both parts together to create an icosahedron.

- Imagine a planar, regular *hexagon* made of cardboard lying flat on your desk in front of you. . . .

- Now draw lines that connect your hexagon's opposite *vertices* and divide it into six equilateral triangles. . . .

- Use a knife to score along the edges of the six triangles. Cut out *one* of the triangles and put it aside. . . .

- Now look at the rest of your figure made of five triangles and *glue* together the two new edges you created by removing the sixth triangle. This causes the figure to *bulge* into a kind of "bowl" formed by five equilateral triangles. . . .

- Now *attach* one equilateral triangle of exactly the same size as the one you cut out before to each of your bowl's five free edges. Now it has a *crown* with five zigzag teeth. . . .

- Construct a second, identical bowl with a *zigzag crown*. . . .

- Place one bowl over the other like a *lid* and rotate it slightly so that the two zigzag crowns *interlock*. Glue each tooth of the lower bowl to *two teeth* of the upper one. . . .

- What does the solid you created look like? . . . How many faces does it have? . . . How many vertices? . . . How many edges meet at each vertex? . . .

- What did you imagine during this exercise in Mathematical Imagining?

Notes on intended mental images

The instructions directly invite students to construct visual mental images. However, they may also produce tactile mental images, and these can accompany and support the imagination process. For example, in order to join the edges created by cutting out the triangle, students need to use a little force to pull the cardboard into shape. This resistance causes the figure to "escape" the plane and move into three dimensions. Students might perceive this tactilely.

The purpose of the first mathematical question, "What does the solid you created look like?" isn't intended to prompt students to give the specific mathematical term (which would not be very useful because they either know it already or not.) Instead, it seeks to elicit an overall picture of the solid so that students visualize it in its entirety. The questions about the number of faces, vertices, and edges are concerned with exploring the solid's combinatorial properties. While the construction process makes it possible for students to name the number of faces without having to imagine the entire icosahedron, they can't answer the one about the number of edges per vertex without having at least a mental image of a "bowl." Finally, when it comes to the number of vertices, it is advantageous (but not absolutely necessary) if students can imagine the solid as a whole.

Notes on productive and counterproductive mental images that learners construct

During the process of constructing their mental images, students may encounter and struggle with mental hurdles. Typical counterproductive mental images include the following:

- If they imagine that one bowl is above the other and subject to gravity, then the teeth will hang down vertically. This causes the teeth to not perfectly fit into the gaps between the teeth on the lower bowl (which accordingly, point vertically upward). Instead, the teeth get "wedged" together and don't properly interlock.

- Certain learners notice that the bowl composed of five regular triangles is flexible and not rigid, because two consecutive triangles can function as a hinge. It can then happen that the bowl with its attached teeth has too much give and take. This makes the shape unstable, and students then don't succeed in imagining it as a whole solid.

- Students often imagine the solid is more compressed or sometimes more elongated than a sphere. There are two possible reasons for this. One is that the visualized component parts of the solid—the equilateral triangles—tend to be either too tall or not tall enough (i.e., isosceles instead of equilateral). The second is that its make-up as two congruent components—the bowls—can cause students to focus on the fact that the solid is made of two parts. The result is that they do not perceive the extension of the figure in each of the three dimensions equally. Instead of six fivefold rotation axes, they notice only one. Moreover, they identify two different equivalence classes of vertex neighborhoods instead of just one: the vertices at the North and South Poles, and the vertices at the upper and lower equator.[1]

During this task, students also often find and develop ways of helping themselves overcome the obstacles mentioned here. Typical productive mental images they come up with include the following:

---

1   A solid's rotation axis is *n-fold*, when n is the largest natural number for which a rotation about the axis by an angle of $360°/n$ returns the solid to its original position.

- Students find they can use their fingers to facilitate the construction of the icosahedron. They move the fingers on one hand toward the ones on the other and interleave them—just as the points of the two bowls are brought together and interlocked! This tactile mental image is not intentionally evoked by the instructions. (Just as there is no direct mention of the way in which the pulling force, which may be imagined as a tactile sensation, causes the plane figure to bulge into a bowl.) Nevertheless, it can be developed as a core idea behind the construction of the icosahedron and thus also help make the solid cognitively available for recall later.

- If students visualize the "crown" teeth as being movable, they can adjust them to fit. So even if the teeth on one bowl's crown do not immediately slot perfectly into the teeth of the other one, students can set this right by adjusting the height of the teeth.

- Although the nature of the step-by-step instructions gives a fairly clear indication of the solid's proportions, it may be easier when students imagine an icosahedron that is larger than a fist. Students sometimes say that their icosahedron's diameter is about the same as their own body's. This probably makes it easier for them to explore the solid in detail. In order to answer the question about its combinatorial properties, these students turn the solid around in front of them corner by corner, or even "fly" around their large icosahedron.

Mathematical follow-up questions

This imagining task triggers several mathematical processes, such as reasoning and proving.

- Students can answer the questions: "*How many faces does it have? How many vertices? How many edges meet at each vertex?*" even if they have identified two different congruence classes of vertex neighborhoods. Their answers do not depend on the solid's size or proportions.

- The question, "*What does the solid you created look like?*" is much broader than the questions about the number of faces, vertices, and edges. The course your class discussion takes will depend on whether your students are already familiar with icosahedra. Moreover, the question implicitly assumes that the constructed figure is closed and rigid, although this is not immediately clear from the imagining task's instructions. Why do the teeth of the two bowls fit seamlessly into each other without any gaps?

- One argument that could help during their quest to find answers could go like this: Because the teeth of the crown are triangular, not rectangular, the orthogonal projection of the two perimeters of the bowls (in the direction of the rotation axis) produces two concentric regular pentagons slightly rotated against each other. (Figure 5.1 may help us visualize this argument because we can see the two pentagons, one with solid lines and one with dashed.) Since the projected vertices of one pentagon lie outside the edges of the other and the triangular teeth are attached to a bowl along its edges, each zigzag crown (shaded gray in Figure 5.1) is slightly "open." In particular, none of the planes generated by a tooth are parallel to the rotation axis of the two bowls; that is, the corresponding mental image is not productive and needs further elaboration.

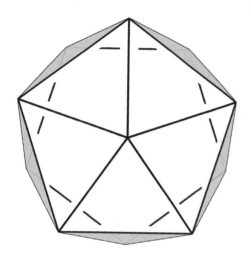

FIGURE 5.1 Slightly open, pointed crown of an icosahedron

- That said, this argument doesn't demonstrate why the two zigzag crowns actually do fit seamlessly into each other or why the solid created is rigid, with equidistant vertices.[2]

This exercise in Mathematical Imagining also triggers processes of varying and generalizing.

- The number of vertices (v), edges (e), and faces (f) are related to one another in an important way, as given by *Euler's polyhedron formula* $v - e + f = 2$. Students can use their answer obtained through counting to verify the formula.[3]

---

2   According to Cauchy's rigidity theorem, two convex polyhedra with the same combinatorics are congruent if their corresponding faces are congruent to each other. In other words, convex polyhedra as solids are already clearly defined through the form and arrangement of their faces. The solid of the icosahedron is not flexible, and its proportions could be said to "take care of themselves" when congruent triangles are joined together as described (Cromwell 1997, 228–233).

3   Euler's polyhedron formula applies not only to convex solids but also to concave ("dented") ones—as long as these do not have any holes and all faces are polygons.

- Conversely, this formula means that the number of edges is already determined when the number of faces and vertices is known. Furthermore, because $q \cdot v = 2 \cdot e$ (where q is the number of faces that meet at a vertex) gives the equation $v - \frac{q}{2} \cdot v + f = 2$, the number of all triangles and vertices also determines the number of triangles that meet at a vertex ($5 = q$). For further details, see exercise C10.

- You might invite students to continue imagining by varying the construction rules. For instance, students may investigate whether using other basic shapes as component parts (regular pentagons, isosceles triangles) or attaching them differently (with not five, but four, or even seven triangles meeting at each vertex) likewise produces closed solids. What do these objects look like?

- Once students have formed a clear mental image of the icosahedron, they can draw on it in subsequent exercises. How can you cut an icosahedron in half to produce a cross section with the shape of a regular decagon? Alternatively, students can investigate the silhouette when the solid is horizontally projected, from "above" (i.e., along projection rays that are parallel to the naturally given, fivefold rotation axis; see Figure 5.1). An even more challenging question is looking at the solid's silhouette under orthographic projection from the "side" (i.e., perpendicular to the rotation axis).

- Students can construct other geometrical solids on the basis of the icosahedron, for instance, its dual, the dodecahedron: "Imagine an icosahedron. First, focus on two of its triangles that are adjacent. Now, join their midpoints with a line segment . . ." (see the comments to C4 and compare with C10).

- The shape of most soccer balls is based on a truncated icosahedron, so you can create an imagining task similar to this one that guides students to mentally construct their own ball. Students should not picture the truncation process as one where they start at the corners and whittle away thin layers. Instead, learners can imagine dynamically varying the position where they cut off the corners and observe how this affects the cross section. This is a typical mental action in the form of a fluid movement (see the comments to C4).

This task naturally lends itself to follow-up questions about different icosahedron sizes. For instance, using coordinate geometry, the equation $A = 5 \cdot \sqrt{3} \cdot a^2$ is obtained for the surface area and $V = \frac{5}{12} \cdot \left(3 + \sqrt{5}\right) \cdot a^3$ for the volume (with an edge length of a).

C2   CROSS-SECTIONING A CUBE

DIFFICULTY LEVEL:	★ ★ ★
MATHEMATICAL IDEA:	Cross sections of a cube taken perpendicular to a space diagonal are regular triangles or hexagons.
PREREQUISITES:	• Position a cube with one space diagonal perpendicular to the desk surface.
	• Distinguish between a solid cube and a wire-frame cube.

Because of its simple, right-angled construction, the cube is one of the most commonly used geometric solids in math lessons. Most of the time, we present a cube in class in a stable position, resting on one of its faces, which is why students usually imagine it in this position. This positioning is conducive for work on many mathematical questions; for others, though, it is not very helpful. For example, if you want to work with triangular plane cross sections or the outline of the cube's isometric projection,[4] it's worthwhile standing the cube on one of its corners.

During this exercise in Mathematical Imagining, students are also going to have some "arts-and-crafts time" in the sense that you instruct them to form a cube out of a lump of clay (or, alternatively, floral foam). Each student positions their cube so that one of its space diagonals is perpendicular to the surface of their desk. Then students cut off the highest cube corner with a cutting plane perpendicular to the chosen space diagonal. Further cross sections taken parallel to the first one produces a series of cross sections in the shapes of triangles and hexagons. To put it another way, students act on their original mental image and keep modifying their cube to obtain a series of mental images: cube cross sections.

- Imagine you have a lump of modeling clay on your desk. . . .

- Now out of that clay, form a cube with height, width, and length that are about the width of your palm, and then put your cube on your desk. . . .

- Now turn the cube so that it is standing on one of its *corners* and hold it in place with one of your hands. . . .

---

4 An *isometric projection* is an orthographic projection along the line joining the origin with the point $(1, 1, 1)$.

- Spatially opposite the corner your cube is standing on, you see another corner. Use your hand to align your cube, so that this *opposite corner* is lying more or less directly above the first corner. . . .

- Now look carefully at this upper corner. There are *three* cube edges that run downward from here. . . .

- Use your other hand to take a knife. Hold it parallel to the surface of your desk and move the knife to the uppermost corner of your cube. Now slice straight through the cube here with a horizontal plane. The shape of this cross section is . . . a small triangle. . . .

- Now slice *further* cross sections by moving the cutting plane downward. . . .

- What shape do your cross sections have? . . . What shape do they have when you keep on cross-sectioning and pass through cube vertices in this process? . . .

- What did you imagine during this imagining task?

Notes on intended mental images

The instructions to this exercise in Mathematical Imagining begin with the mental image of a lump of modeling clay and then guide students to work with it in a series of steps. Students form a cube and then use mental manipulations to take many thin cross sections from it. Continuous, fluid mental movements such as morphing are not specifically called for in the task, even though students will construct such mental images, and they may even prove to be productive. The intended mental images are all of a visual nature; however, the first contact with the construction material might be accompanied with tactile mental images of moist or cool clay.

The mathematical questions: "What shape do your cross sections have? What shape do they have when you keep on cross-sectioning and pass through cube vertices in the process?" are designed to help students visualize the shape their series of cross sections will have depending on where they slice the cube. Cross sections taken from the upper and lower thirds of the space diagonal are equilateral triangles, while those from the middle third are hexagons. If the cross section is taken with a plane slice through the cube's center, it will be a regular hexagon due to symmetry. This figure coincides with the outline of the cube's isometric projection.

Notes on productive and counterproductive mental images that learners construct

When working with this exercise in Mathematical Imagining, students repeatedly come up with productive mental images.

- To answer the mathematical question, students may find it advantageous to change the cube's position from that given in the instructions or to vary their vantage points. If their line of sight is, for instance, along the space diagonal, the cube's threefold symmetry emerges particularly well.

- When students change their line of sight while observing their cube, they may also reposition it, for example, place it on a slightly open palm. Should they then be able to "feel" the clay's coolness or weight, then their according mental images are more haptic instead of visual.

- Instead of placing their cube on a surface, some students suspend it instead and hang it by one corner. The cube then, thanks to gravity, aligns itself so that the corresponding space diagonal is perpendicular to the desk surface.

- Students can and do supplement or even partially replace their mental image of the solid cube with its wire-frame model: from the uppermost corner, three edges run symmetrically downward, lead to three additional corners of the cube, and then each forks into two new edges, and so forth.

- The occasional student will utilize liquid ink or paint in their mental images and let it run down from the upper cube corner slowly and evenly along the edges. They concentrate on the points where the paint is advancing and use these endpoints to create their individual cross sections.

- Instead of taking multiple, individual cross sections, students can construct a series of mental images that morph seamlessly into the next image, like in an animated film. First, a small triangle appears and starts getting bigger. After a while, the expanding triangle's corners break off—and new sides grow in their place. In this way, the triangle morphs into an (irregular) hexagon. As the film continues, the shorter sides increase in size at the expense of the larger ones until, eventually, all the sides have the same size and a regular hexagon appears. Then the entire process is repeated in reverse order. Hence, one sees that all the individual cross sections are the metamorphosis of a triangle.

Some counterproductive mental images that students tend to construct during this imagining task include the following:

- The supple nature of the clay means that the cube constructed from it can be easily deformed, which leads to students having mental images of the corner that the cube is resting on is somewhat caved in.

- If a student imagines the orthographic projection of the cube parallel to the space diagonal, he will obtain a hexagonal silhouette. This mental image can become so dominant that it blocks the mental image of the hexagonal cross section obtained by projecting in the direction of the space diagonal. (The two hexagons are rotated by 30° with respect to each other.) Presumably, a Gestalt principle is at work here.

- Many cubes mutate when students work on them from the point of the upper vertex. When students concentrate exclusively on the edges originating here, they block out the rest of the cube. Then, in the moment when they want to visualize it again, they suddenly have the mental image of a pyramid (or, in rare cases, even a cone). The reason for this might be that they have learned to associate a point where edges come together with pyramids and cones, and so they interpret this characteristic as being a typical feature of those geometrical shapes.

- Some students do not manage to keep their cubes in the described position; their cubes tend to fall over and land in the stable position, lying on a face. In this standard way of looking at cubes, the fourfold symmetry of the visible faces is dominant; hence, the corresponding rotation axes seem to be the "natural" line of sight. Indeed, the fourfold symmetry can be so suggestive that the imagined cross sections turn out to be quadrangular!

Mathematical follow-up questions

You can illustrate the cross sections at the heart of this task by filling a Plexiglass cube with colored water and then positioning it on one of its corners. As soon as the water drains away, its surface will change shape and reveal the metamorphosis of the cross sections.

This exercise in Mathematical Imagining initiates various mathematical processes, such as varying and conjecturing. After completing this task, related follow-up questions can be asked. Students can explore many of them imaginatively.

- The cross sections you took during the imagining task exhibit three or six sides. Can you slice straight through a cube so that the cross sections have four or five sides? Can there be cross sections from a cube with seven or more sides?

- Which (Platonic or Archimedean) solids result when you slice off all the corners of a cube equally?

- The cut that produces a cross section with the shape of a regular hexagon also cuts the cube into two congruent halves. What other cuts also bisect the cube like this? (See also the end of this exercise.)

- After a while, students tend to assume that the regular hexagonal cross section has a larger area than all the other cross sections obtained by cutting with parallel planes. How do you slice the cube to get a cross section with an even larger area? How must you slice the cube in order to produce the cross section with the maximal area?[5]

- The equation $V_{\text{pyramid}} = \frac{1}{3} \cdot V_{\text{cube}}$ raises the question of if and how a cube can be dissected into three congruent, nonright pyramids. Here, as well, the atypical way of positioning the cube balanced on one of its vertices can help students to imagine this situation. Each pyramid's base builds on one of the cube's lower faces while the pyramid's apexes come together at the cube's uppermost vertex. All three pyramids touch at the cube's diagonal. When the cube is placed in its standard position on one of its faces, though, it would be much more difficult to imagine the three pyramids inside the cube.

---

5   The maximal cross-sectional area results by cutting a cube with a plane through two diagonally opposite edges and along two opposite face diagonals. For a silent, animated geometry film showing sectioning a cube in many different ways, see https://www.atm.org.uk/Geometry-and-Tessellation-interactive-films.

- The threefold rotation axis can be seen in the context of the symmetry group of the cube: How many threefold rotation axes can you find? And which other rotation axes (twofold, fourfold, . . .) does the cube have?

- How large is the area of the cross section when the three faces of a cutoff cube corner are equal sized and still triangular? Based on De Gua's theorem—a spatial generalization of the Pythagorean theorem—the area must be $\sqrt{3}$ times as large as one face (see comment to R10).

- What does the solid look like that results from rotating the cube about a space diagonal? As the family of generating lines in the upper and in the lower part intersect the rotation axis, two cones result here. In the middle section, a one-sheeted *hyperboloid* is formed, because here, the family of generating lines and the rotation axis are skew (see Figure 5.2). Its volume is $V = \dfrac{\sqrt{3}}{3} \cdot \pi \cdot a^3$ (with edge length a).

You can also connect this exercise in Mathematical Imagining to aspects of architecture and art.

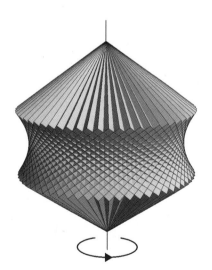

FIGURE 5.2 A cube rotated about a space diagonal

- For numerous practical reasons, houses are built so that the walls of their rooms are perpendicular to their floors. One exception to this practice are so-called cube houses that were constructed in the Netherlands in the cities of Helmond (1974–75) and Rotterdam (1982–84) and designed by the Dutch architect Piet Blom. These houses consist of a cube placed on one of its corners and supported by a pylon positioned in the direction of the cube's space diagonal. These houses have three floors: the middle one is a regular hexagon, while the first and third floors are regular triangles.

- Artistically, the Swiss artist and designer Max Bill worked with bisecting cubes. For his installation *schtatt e schtadt e schtatt* (1977), in an urban park in Jerusalem, he bisected solid cubes made of black granite in four different ways. Among others, he thus created a stone sculpture with a regular hexagonal cross section. What could the other three look like?

C3   LAYING OUT A HARMONIC PYTHAGOREAN TRIANGLE

DIFFICULTY LEVEL: ★

MATHEMATICAL IDEA: Laying out a triangular dot pattern with ten checkers pieces.

PREREQUISITES: • Place checkers game pieces in a regular pattern.

The triangular tetractys pattern of dots is mostly taught at school in the context of triangular numbers. This means most students rarely learn that the tetractys stems from Pythagoras and had great symbolic significance for him and his followers. This is actually a pity because, after all, when we judge whether a number is special or even beautiful, we are guided by similar principles that once influenced Pythagoras.

This exercise in Mathematical Imagining asks learners to lay out a number of checkers in rows that create a dot pattern in the shape of regular triangles. This leads learners to triangular numbers.

- Imagine you have a pile of checkers. Take *one* checker and place it on your desk in front of you. . . .

- Take *two* more checkers. Place one of them *beneath and to the left* of the first checker you put on your desk and the other one beneath and *to the right* of your first checker. These two checkers now lie in a *horizontal row* beneath the first checker. Do you see the pattern the three checkers form? . . .

- Now, pick up another *three* checkers and once again place them *beneath* the pattern you created, and space them nice and evenly in a horizontal row. . . .

- Finally, take *four* checkers and lay them out in a row beneath the pattern. . . . This creates a dot pattern made up of four rows of dots.

- What shape does this pattern have? . . . How many checkers make up the pattern? . . .

- What did you imagine during this exercise in Mathematical Imagining?

Comments

The sum of the first four natural numbers is ten. Because of the triangular shape of the pattern, the number of checkers is what is known as a triangular number. In other words, ten is the fourth triangular number.

Mathematical follow-up questions

What, then, is the fifth, sixth, ..., hundredth, ..., nth triangular number? If students continue the process described in the exercise in Mathematical Imagining row by row, then, at some point, some of them will most likely wonder how many checkers are necessary up to the nth row. This imagining task can then be used to derive *Gauss's formula* to find the nth triangular number. The triangular arrangement of $1 + 2 + 3 + ... + n$ and its copy can be arranged as a parallelogram (see Figure 5.3). For the number of dots, it thus follows that $1 + 2 + 3 + ... + n = \frac{1}{2} \cdot n \cdot (n + 1)$.

FIGURE 5.3
Visual proof of Gauss's formula

This summation formula is perhaps the most important example of an *arithmetic progression* (for an imagining task on the sequence of consecutive odd numbers, see PS5).[6]

The Pythagoreans, a religious and political society in ancient Greece led by Pythagoras, closely studied numbers related to regular geometric figures (*figurate numbers*). As such, the equilateral triangle composed of four rows with a total of ten pieces or dots was a central symbol for the Pythagoreans. They saw this figure—the *tetractys*—as representing fundamental aspects of their worldview. The number four, for instance, stood for that which was perfect and harmonious: the Greeks identified four seasons, four elements, and four temperaments. The chord ratios of important musical harmonies are based on the first four numbers: 1:2 (octave), 2:3 (fifth), and 3:4 (fourth). Further, the Pythagoreans held the number ten to be a perfect one since it formed the basis of the number system. In the tetractys, ten is the sum of the first four numbers, so to the Pythagoreans, the tetrad had "begat" the number ten.

The harmonic triangle thus symbolizes the Pythagorean creed "Everything is number"; that is, everything that is harmonious and beautiful in nature can be described with natural numbers or with ratios between them. The Pythagoreans' philosophy could also be put more generally: significant phenomena can be

---

6   For another visual proof of Gauss's formula, see Nelsen (1993, 70).

described with "simple" numbers (i.e., whole and rational numbers), and, conversely, simple numbers intimate significant things. This is why the irrationality of numbers such as $\sqrt{2}$ was so shocking for them.

Don't these Pythagorean views also underlie the way we see numbers today? We celebrate birthdays with the passing of each year of a person's life. Why don't we celebrate every 197 days? Moreover, we consider birthdays that conclude a period of ten years to be particularly significant, and in certain languages, these anniversaries are even invested with geometric, aesthetic notions like "round" (for those ending in a zero). Prices are another example of this phenomenon. We notice if a supermarket purchase costs "exactly" one hundred dollars, or at least we are more likely to notice a round price like this than a purchase of exactly nineteen dollars and seven cents. But why exactly is this so?

This same Pythagorean approach of seeing simple numbers and proportions as signs of significant phenomena is also at the root of questions that arise when we try to make mathematical connections. For instance, the ratio $1:3$ in the formula for the volume of pyramids ($V_{\text{pyramid}} = \frac{1}{3} \cdot Bh$ or $V_{\text{pyramid}}:(Bh) = 1:3$) may prompt the geometric conjecture that any three congruent pyramids can be joined together to form a parallelepiped or cuboid. Under what conditions does this apply?[7] (For sectioning a cube into three congruent pyramids, see exercise C2.)

Wagenschein also addresses the Pythagorean in us by stressing a specific feature of the surface area of a sphere. We perceive the full moon in the sky as a circular disk ($A_{\text{circle}} = \pi r^2$), even though we are seeing a hemisphere ($S_{\text{hemisphere}} = 2 \cdot \pi r^2$). Who would not wonder whether there might not be a deeper meaning worth uncovering behind the equation $S_{\text{hemisphere}}:A_{\text{circle}} = 2:1$?

One last example is from integral calculus. If we read the indefinite integral $\int x^2 \, dx = \frac{x^3}{3}$, as a ratio equation $\int x^2 \, dx : x^3 = 1:3$ and the term x^3 as the volume of a cube, we could ask ourselves the following Pythagorean-styled questions (see also my comment in C2): Why is $\int x^2 \, dx$ exactly one-third of a cube? What does this integral have to do with the volume of a pyramid? You can see this when you take x^2, not as the length of a line segment but as the area of a square. If x moves from 0 to 1, instead of interpreting the integral $\int_0^1 x^2 \, dx$ to geometrically sum up

---

7   The question connects to Hilbert's third problem that Max Dehn solved in 1901. As a consequence, not every pyramid can be cut into finitely many polyhedrons and then skillfully rearranged and joined to form a cube with equal volume. The general formula for calculating the pyramid volume cannot be proved using elementary "cut-and-glue" methods but requires infinite processes (such as Cavalieri's principle, or integral calculus). This is in contrast to the two-dimensional space where every triangle can be bisected and rearranged into a parallelogram—and with one cut more, even into a rectangle).

the line segments, view it as summing up the squares x^2. When these (infinitely many) squares are combined to a solid, they are just cross sections of that solid. Its boundary curve is a straight line ($y = x$), which means that all the sections together form an oblique, square-based pyramid. To sum up, the ratio of its volume to the surrounding unit cube is $1:3$, which in the case of the traditional interpretation of $\int_0^1 x^2 \, \mathrm{d}x$ means that the area under the parabola $y = x^2$ and the surrounding unit square are in ratio of $1:3$.

C4   TRUNCATING A TRIANGLE

DIFFICULTY LEVEL: ★

MATHEMATICAL IDEA: The process of cutting off the corners of a triangle ever more deeply results in a series of hexagons and, finally, a triangle.

PREREQUISITES:

- Familiarity with the shape of an equilateral triangle.

- A trapezoid is obtained by cutting the corner of a triangle off with a line parallel to its opposite side.

Truncating solids can be a particularly productive process. For example, you obtain an octahedron when you cut off the corners of a tetrahedron. My recommendation is for learners to experience this procedure with plane figures first.

In this exercise in Mathematical Imagining, students truncate a paper triangle by cutting off its three corners. First, they cut off only tiny bits of the corners and then cut more extensively. In this way, they obtain a series of nested hexagons starting from one triangle and ending in the (midpoint) triangle.

- Imagine you have a piece of paper and a pair of scissors. . . . Cut an *equilateral triangle* out of your paper. Turn your triangle so that one of its sides is horizontal and is your triangle's *base*. Now the corner of the opposite side is pointing *up* and forms the top of the figure. . . .

- Concentrating now on the *top* of your triangle, pick up your pair of scissors and bring them close to the triangle's top. Hold your scissors parallel to the bottom side of the triangle and cut off a tiny bit of the top. . . . This gives your original triangular figure a new *side* . . . and results in a trapezoid-shaped quadrangle: a triangle with a *cutoff* corner. . . .

- At the bottom left and right, you see the two other corners of your original triangle. First, cut off the *same small amount* of the corner on the *left* like you did before with a cut parallel to its *opposite* side. . . . Now, cut off the corner to the *right*, again in the *same small amount* and with a cut that is parallel to the side opposite the *right* corner. . . . In this way, you now have a paper figure that looks like a triangle with all corners cut off. . . .

- How many corners does your new figure have? . . .

- Now concentrate on the three *sides* of your figure that appeared when you cut off the triangle's corners. Cut *further small* strips away from these sides. . . . What kind of *shape* does your paper figure take on? What shape would it have if *you kept* on cutting away such small strips? . . .

- Keep cutting like this *only until* two cut edges meet at one of the midpoints of your original triangle's sides. . . . What shape does your paper figure have? . . .

- What did you imagine during this exercise in Mathematical Imagining?

Comments

When students cut off the three corners of the original figure, they produce a hexagon. When they keep trimming symmetrically, they get smaller and smaller hexagons that lead to a triangle. This figure stands on its apex and, therefore, rotated 60° when compared to the original figure. As the midpoint triangle, this new figure divides the original triangle into four congruent triangles that, therefore, all have the same area.

Just like with the cube cross sections in C2, students can construct a wide variety of individual shapes in the form of an animated film composed of a morphing mental image. First, the animated film shows a triangle. From its corners, new sides begin to grow that make the old triangle turn into an irregular hexagon. As the film continues, the sides of the original figure keep getting trimmed, causing them to shrink. Then, for a brief moment, the original sides have the same length as the newly created cut edges and then become increasingly shorter. Finally, the original sides remain only as points, and these are the vertices in the resulting final figure of an upside-down triangle. In this way, you get a different triangle metamorphosis than with the cube cross sections of C2.

Mathematical follow-up questions

You can vary the question raised in this imagining task in many ways. Here are some questions you can ask your students: What changes happen to your figure, when instead of starting with an equilateral triangle, you truncate a right-angled isosceles triangle—or even a general one? What sequence of cross sections do you get when you cut off the corners of a square or an irregular quadrilateral? Questions like these could also be the focus of further exercises in mental imagining. You can also transfer them to solids: What solid results when, for example, you cut off the four corners of a tetrahedron and then keep trimming the newly created cut surfaces? At first, a truncated tetrahedron results, which is an Archimedean solid. It

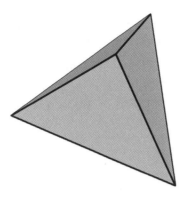 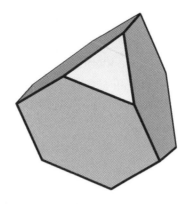 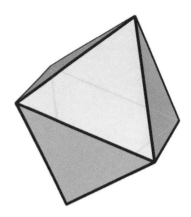

FIGURE 5.4   A sequence of progressively truncated tetrahedrons

consists of four regular triangular faces (the cut surfaces of the four cutoff corners, depicted in light gray in Figure 5.4) and four regular hexagonal faces (the thereby trimmed triangular faces of the original solid, in dark gray).

If you continue truncating, this solid then transforms into an octahedron as follows. The four cut surfaces remain triangular, while the further trimmed faces of the solid turn into triangles (in dark gray). Should the original tetrahedron be truncated even beyond the octahedron, a solid in the shape of a tetrahedron results again. Its four corners are—similar to the midpoint triangle in the imagining task—at the midpoints of the original tetrahedron's faces. This is why tetrahedrons are *self-dual* (compare to the comment in C8).

Your more experienced classes can even truncate an icosahedron, first just a little, and then become more ambitious with their cuts (based on the imagining task C1). In this way, a shape emerges in which each pentagonal face will be surrounded by five hexagons. This geometry—a truncated icosahedron—is the same as the structure of soccer balls, which can be sewn out of twelve pentagonal and twenty hexagonal leather pieces.[8] It is also the same structure of a new crystalline manifestation of carbon, a macromolecule composed of twelve pentagonal and twenty hexagonal carbon rings (so-called *fullerene* or *buckyball*). If truncation of the original icosahedron continues, first an *icosidodecahedron* (where every pentagon is surrounded by five triangles) emerges, followed by a dodecahedron. Students wanting to construct this challenging series of mental images can be supported in the task through the use of appropriate visualization software.[9]

---

8   For more geometry of classic and modern soccer balls, see Glaeser and Polthier (2020).

9   With the program KaleidoTile, students can truncate Platonic solids interactively (http://www.geometrygames .org/KaleidoTile/index.html).

Instead of truncating the corners of a Platonic solid, students can also cut off its edges. Here, again, convex solids result, which are composed of more than one type of regular polygon and have identical vertices, meaning that the arrangements of polygons around each vertex are identical (known as Archimedean solids). However, there are Archimedean solids that can't be formed by truncating a Platonic solid. Which ones are they?[10]

---

10   Seven Archimedean solids can be formed through truncating the corners of a Platonic solid, and two Archimedean solids result from truncating the corners and edges of a Platonic solid: the (small) rhombicuboctahedron and the (small) rhombicosidodecahedron (Cromwell 1997, 79–85).

C5   CONSTRUCTING A PENTAGRAM

DIFFICULTY LEVEL: ★

MATHEMATICAL IDEA: A pentagram can be constructed by following combinatorial instructions.

PREREQUISITES:
- Properties of a circle.
- Jumping and skipping while playing board games (e.g., Chinese checkers).

A pentagram and a pentagon are related in many ways. For example, a pentagram can be constructed from a pentagon by extending all its sides until every pair of sides intersects. Or it can be constructed by drawing all the diagonals of a pentagon.

This exercise in Mathematical Imagining doesn't make use of the geometrical commonalities between a pentagon and a pentagram in order to construct the latter but, instead, takes a combinatorial (graph-theoretical) approach. For the pentagon, the five vertices are arranged uniformly on a circle and connected consecutively. For a pentagram, though, on each step, you skip over the next vertex and connect to the following one.

- Imagine a circle. . . . Now put a mark at the *top* of your circle's circumference in the same place where there is a 12 on a clock's face. . . .

- Now mark four *more* points on the circumference. Arrange them so that, in the end, all five points are *equally* spread around the circumference. Do you see the circle with all five points? . . .

- Now focus on the first point at the top of the circle. From here, jump over the point right next to it and land on the point *after that*. Keep jumping like this, always skipping over one point to reach the one next to the point you skipped over. Continue doing this until you come back to the point where you started. . . . You will reach your starting point after you have gone around your circle *twice*. . . .

- After you successfully finish this step, draw line segments to *connect* the points—in the exact order that you landed on them by always skipping over one point. . . .

- What figure results? . . . How many line segments does it have? . . .

- What did you imagine during this exercise in Mathematical Imagining?

Comments

A star-shaped figure with five points is called a *pentagram* or a *regular star pentagon*. Its five sides all intersect each other, and it can be drawn in one fell swoop, without having to lift the pen from the paper. It doesn't matter whether your students jump around their circle in a clockwise or counterclockwise direction, they will always generate the same pentagram.

In reference to drawings and sketches from the Middle Ages, in which pentagons were drawn on the human body so as to connect the human physical shape to the golden ratio, students could imagine the five points as the endpoints of their own body with outstretched arms and slightly spread legs. With this mental image in the background, it might be easier for them to have a good picture of the points in order to connect them properly. Especially younger students like to try the hands-on approach and enjoy using yarn or elastic bands to construct the pentagram onto a circular geoboard with nails arranged in a circle.

Mathematical follow-up questions

This exercise in Mathematical Imagining has scope for a wide range of questions and discoveries about the pentagram:

- Working with questions about the interior angles could be approached as follows: If the points where the five sides intersect[11] are not considered to be vertices and the pentagram is traversed along its five edges, the sum of the interior angles of the five corner angles is $180°$ (see imagining task R1). If, however, the figure is considered as a concave decagon (by ignoring, for example, the five short inner segments), then the five new interior angles yield $5 \cdot 360° - 540°$ more, resulting in $4 \cdot 360°$. No matter how you look at it, the formula $(n-2) \cdot 180°$ doesn't apply to self-intersecting polygons.

- If we focus on two intersecting sides of the pentagram and restrict attention to these two sides alone, the question becomes, What is the ratio of the longer side segment to the shorter one, or what turns out to be the same, What is the ratio of the entire side to the longer segment? Typically, this question is answered with help from similarity and proportion equations. When the shorter side segment has the length 1 and the longer segment's

---

11 We say two sides *intersect* if they share a point that is not an endpoint of either side. A polygon with sides that intersect is said to be *self-intersecting*.

is φ, then the equation $\frac{\varphi}{1} = \frac{\varphi+1}{\varphi}$ leads to a quadratic equation that has a segment ratio equal to those of the *golden ratio,* $\varphi = \frac{\sqrt{5}+1}{2} \approx 1.6180$.

- Historically, it seems that the existence of irrational numbers was first discovered via the golden ratio and not via the diagonal of a square. Legend has it that this discovery was made by a Pythagorean and that, by doing so, he broke the Pythagorean's central precept of "Everything is number" and paid for this "crime" with his life (see also my comment to imagining task C3).

- The irrationality of the ratio φ can be proved with a geometric procedure that, with good reason, reminds us of the geometric Euclidean algorithm (with its mutual subtraction) for finding the greatest common measure (*anthyphairesis*). However, before we can carry out this algorithm, we first need to attach a second smaller pentagram to two neighboring vertices of the first pentagram constructed in the imagining task: The smaller pentagram's sides are the same length as the longer segment of the larger pentagram's sides (depicted with dashes in Figure 5.5). The points where each pair of sides intersect on the larger pentagram divide each side into two segments whose lengths are related by the golden ratio, which is also called the *golden section*. This is why one corner of the second, smaller star coincides with one of these points of intersection. As a result, the pentagrams are similar: ratios of corresponding segments are constant. This is why an infinite number of smaller pentagrams can be attached to the larger one.

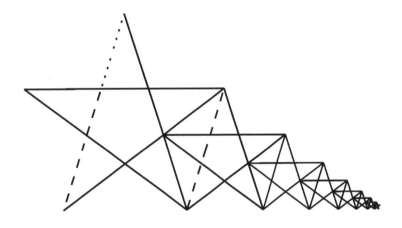

FIGURE 5.5 Geometrical mutual subtraction in a pentagram to prove the irrationality of the golden ratio

- Geometric mutual subtraction is an algorithm that, for any two segments, determines the greatest common segment that they can be measured with, in the sense that the original segments are both integral multiples of this greatest common segment. So, to measure the full side length using the longer (dashed) segment of a side in the largest pentagram, you subtract that segment from the full side and the shorter segment remains (dotted section in Figure 5.5). When you next try to measure the longer segment with the shorter one, you are back at the starting point, at a smaller scale, as a look at the smaller, attached pentagram shows. The longer segment of the larger pentagram becomes the entire side of the smaller one, and the shorter segment of the larger pentagram is then the smaller one's longest segment. As this phenomenon repeats itself with each subtraction, the algorithm never stops. In other words, a full side of the pentagram and its longer subsegment do not have a common, smallest subsegment, which is why the golden ratio cannot be written as a simple fraction.

- The pentagram as a symbol also has cultural meaning. The inverted pentagram comes to mind, as it has been associated throughout history with both "good" (as a symbol used to ward off evil, featured on heraldic coat of arms) and "evil" itself (black magic, satanic cults). The pentagram also appears on the logos of some secret societies (the Pythagoreans, the Freemasons) and automobile manufacturers. Pentagrams are particularly and noticeably present on national flags (and not just those of socialist states!). Pentagrams and five-pointed stars grace the flags of about fifty countries, including Morocco (pentagram) and the United States and China (five-pointed stars).

The combinatorial (graph-theoretical) method used here to construct the pentagram also raises questions that can go far beyond the pentagram itself.

- What kind of figure results when you jump over more than one point and, for example, "land on" and connect every third point?

- What kind of figure do you get when you place six points equally spaced around a circle and connect every second point? Every third point?

- In general, when you place n points equally spaced on a circle and connect every pth point (modulo n) in one direction, you will get (after one or more trips around the center) a star polygon. Following Coxeter, we denote this with the *Schläfli symbol* $\left\{\frac{n}{p}\right\}$ (see also exercises C10 and C13).[12] Here, there are three cases to distinguish.

1. When n and p are relatively prime, a sequence of n line segments arises, forming a connected figure known as a *regular star polygon*.

2. When n and p have a greatest common divisor $\mathrm{GCD}(n,p)$ satisfying $1 < \mathrm{GCD}(n,p) < p$, a star polygon also results; however, one that doesn't connect all n but only $\frac{n}{\mathrm{GCD}(n,p)}$ vertices. In the smallest, nontrivial case $\left\{\frac{10}{4}\right\}$, just five points are joined, resulting in a pentagram.

3. When the greatest common divisor of n and p is p, then the figure closes up after the first revolution around the circle. Consequently, this results in a regular polygon with $\left\{\frac{n}{p}\right\}$ sides and vertices. In the smallest, nontrivial case $\left\{\frac{6}{2}\right\}$, three of the six points are joined, which results in a regular triangle.

In the last two cases, all points on the circle can be included: Just take $\mathrm{GCD}(n,p)$ copies of the polygon and rotate them individually until every point on the circle is covered by a vertex of one of the copies. As a result, you will get a *compound star polygon*.

*   As argued earlier, the sum of the interior angles of a pentagram is $180°$. What is the sum of the interior vertex angles of a star polygon $\left\{\frac{n}{p}\right\}$? With the same argument, you get $n \cdot 180° - p \cdot 360°$, and interestingly enough, the result holds regardless of whether the star polygon is regular or compound.

*   How many different star polygons can be constructed starting from n points equally spaced on a circle? If you only count regular star polygons, there are $\frac{1}{2} \cdot \varphi(n) - 1$ of them. If you include the compound ones, there are $\left\lfloor \frac{n-3}{2} \right\rfloor$ star polygons.[13]

---

12   In the Schläfli symbol, as the integer p must satisfy $p > 1$ and because of symmetry reasons $p < n/2$, the integer n satisfies $n > 5$.

13   The function $\varphi(n)$ denotes the number of positive integers up to a given integer n that are relatively prime to n (also called *Euler's totient function*). The function $\lfloor x \rfloor$ gives the largest integer less than or equal to x (also called *floor function*). In contrast to the way presented here, some authors subsume convex polygons in star polygons, which is why their formulas differ.

C6   FOLDING A TETRAHEDRON

DIFFICULTY LEVEL: ★ ★

MATHEMATICAL IDEA: Tetrahedra can be folded from a rectangular sheet of paper without making cuts or overlaps.

PREREQUISITES: • Rolling a cylinder tube from a sheet of paper.

You can create a customized package for polyhedral objects, such as shoeboxes, by enveloping them in a piece of paper. However, if the paper's shape doesn't correspond to the object's *geometrical net*,[14] the resulting packaging or container may end up with gaps or overlaps that must be sealed shut or cut off.

In this exercise in Mathematical Imagining, your students will fold a tetrahedron out of a piece of rectangular paper without having to make any cuts—and without the result having any gaps or overlaps.

- Imagine you have a piece of standard-sized paper. . . . Now lay it down *horizontally* in front of you on your desk, in landscape orientation. . . .

- Bend your paper by bringing the left and right edges together so they just meet. Now you have a short *cylinder. Tape* together the two edges you joined. . . . Imagine how your cylinder is lying on the desk. The *rear* opening is pointing *away* from you and the *front* one is pointing *toward* you. . . .

- *Close* the front opening by pressing it flat with one of your hands. Hold the flattened edges together and *glue* them together. . . .

- You also need to close the rear opening of your cylinder but a little bit differently: Hold out your hands as if you are about to clap them. . . . Position your hands so that the rear opening of the cylinder is *between* your hands and then bring your hands together. This *presses* the rear opening *together* and it closes. Hold the flattened edges together and *glue* them also together. . . .

- Now you have a paper object that is *closed.* This solid has two *edges*: a rear one and a front one, but these edges are *rotated* in relation to one another by ninety degrees. . . .

- Your solid also has four pointed *corners:* the front *left* and front *right* and the rear *up* and rear *down.* . . . There are four slight, outwardly curved *ridges* in the paper that run between the front and rear corners. . . .

---

14   A *net of a polyhedron* is a polygon that can be folded up into the polyhedron.

- Now carefully crease the ridges until they become real *edges*. . . . Now you have two creased edges that run from the top of the rear corners to both front corners and just like this, two further creased edges run from the bottom of the rear corners to both front ones. . . .

- What kind of solid do you have? . . .

- What did you imagine during this exercise in Mathematical Imagining?

Comments

The result is a tetrahedron that is slightly stretched out in the direction from front to rear. To achieve a regular tetrahedron, you would need to use a piece of paper with a different size than standard letter size, namely, one whose dimensions exhibit an aspect ratio of $4:\sqrt{3}$ (length to width). In other words, instead of using a typical triangular tetrahedron net, here, you can use a piece of rectangular paper to create a faceted model of a regular tetrahedron.

All other Platonic solids can also be folded with rectangular paper. However, it seems to be unavoidable that the solids' surfaces will inevitably have places that are covered with more than one layer of paper.

Mathematical follow-up questions

One follow-up activity that can take you far is the following: Glue one face of a regular octahedron to one face of a regular tetrahedron. Both figures have the same edge length. How many edges, corners, and faces does the new solid have?

Students might be tempted to merely subtract the two joined faces from the sum of both solids' faces, as well as the three (from six) glued-together edges. However, each pair of triangles that has a shared edge created by gluing the figures together is in the same plane. Hence, the adjacent triangles are coplanar and form a single rhombus. As a consequence, each such pair of coplanar faces counts only as one face, and the edge between does not count. You can help your students see this, for example, by calculating the angle between two consecutive faces of each solid, the dihedral angle. By trigonometry or coordinate geometry, the dihedral angle can be shown to be $\arctan\left(2\sqrt{2}\right) = \arccos\left(\frac{1}{3}\right) \approx 70.53°$ for the tetrahedron and $\arccos\left(-\frac{1}{3}\right) \approx 109.47°$ for the octahedron, which sum up to $180°$. But you can also demonstrate the coplanarity through a more direct argument. First, remember that a regular octahedron is a regular tetrahedron with three smaller tetrahedrons removed (see Figure 5.4). Now, if one of these small tetrahedrons is rejoined to the octahedron, every nonglued tetrahedron face must inevitably be coplanar with an octahedron face. As a consequence, the combination of an octahedron and a tetrahedron is a solid with seven faces (three rhombuses, four triangles), twelve edges, and seven corners.

The question whether these adjacent faces really lie in a plane can also be explored from a different, more global position. To do so requires slightly broadening the task's scope. As is known, three-dimensional space can be completely filled in using only congruent cubes through stacking them next to and on top of each other. This is not possible to do using only congruent regular tetrahedra, despite Aristotle's supposed claim to the contrary. Nevertheless, there is another way to fill space using Platonic solids, based on cube stacking. For this *space-filling tessellation*, four nonneighboring corners of a cube are cut off to produce a tetrahedron (see Figure 5.6). At the same time, each cut-off corner is the eighth part of an octahedron, as these "corners" are in fact three-sided pyramids with right angles at the apex and an equilateral base. When these pyramids are combined with the cutoff corners of the seven other cubes sharing this same apex, a complete octahedron results.

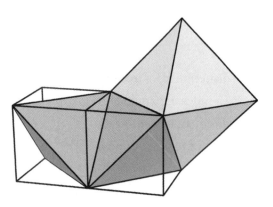

FIGURE 5.6 Filling space with regular tetrahedra and octahedra

In this way, out of the familiar tessellation of space with cubes, we get a different regular tessellation, now with two Platonic solids. The octahedra have the same orientation and stand on a corner, as they are usually depicted. The tetrahedra, though, appear in two different orientations and stand on an edge. In every vertex of the tessellation, eight tetrahedra and six octahedra meet. For reasons of symmetry, the adjacent faces of the tetrahedra and octahedra once again must lie in the same plane.

There are many other convex polyhedra that fill space. For example, take an octahedron and attach two tetrahedra, one on each of two parallel faces of the octahedron. Because several adjacent triangles are coplanar, you get a polyhedron that is a parallelepiped with six congruent rhombic faces (called *rhombohedron*). You can think of it as a solid that results after threefold shearing of a cube. By its construction, this polyhedron also fills space (Figure 5.7).[15]

---

15 For further convex, space-filling polyhedra, see Weisstein (2003, 2760–2761) and Glaeser and Polthier (2020).

For many years, the beverage industry has made use of the efficient construction of tetrahedra (the basis of this imagining task) to make product containers. In this industrial packaging, a beverage-filled carton tube is clamped, sealed, trimmed, conveyed down the assembly line, and rotated by 90° in order to repeat the clamping, sealing, and trimming procedure. This rational production method of tetrahedron-shaped beverage contain- ers has been in place in Sweden since 1952 and led to the brand name Tetra Pak, which has worldwide recognition. However, tetrahedra can't be stacked easily without wasting space, and they also do not fit well in right-angled boxes. For these and other reasons, the Tetra Pak was phased out of the market at the end of the 1970s and replaced by the boxy Tetra Brik.

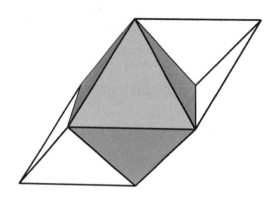

FIGURE 5.7 Regular octahedron with two attached tetrahedra

There is also another reason why tetrahedron-shaped packaging is not optimal. Modern combinatorics describes how you must crease and fold the net of a regular tetrahedron to obtain a solid with a 37 percent greater volume—and do this with the same surface area! In contradiction to the intuitive assumption that a solid with a given surface area automatically has to be convex to enclose maximum volume, the solid that results here has dents in it, meaning it is concave.[16]

---

16   According to a *theorem of Bleecker* (proved in 1996), the net of each Platonic solid can be folded into a poly- hedron with greater volume than the original solid. In the process, though, the solids inevitably become concave. To watch a silent animated film that shows this volume increase for tetrahedra and cubes, see http://www.etudes .ru/en/etudes/polyhedra-volume-increasing.

C7 CONSTRUCTING FIGURES OF CONSTANT WIDTH

DIFFICULTY LEVEL: ★ ★

MATHEMATICAL IDEA: The circle is not the only convex shape of constant width.

PREREQUISITES:
- Construct an equilateral triangle with a compass.
- Construct parallel line segments.

If you measure the width of any convex planar shape,[17] you might get a different length, depending on the direction in which you take the measurement. The circle, however, is a shape with a width that is the same length no matter in which direction you measure it. Contrary to expectation, though, constant width does not guarantee that you have a circle. In fact, infinitely many convex figures are not circular but have constant width: they all can be rolled arbitrarily between two parallel tangent lines.

- Imagine you have a piece of cardboard. Now get ready to draw three points on it. Draw one point on the *left*, . . . and a handbreadth's distance to the *right*, draw the second point. . . .

- To make your *third* point, take a *drawing compass* in your hand. Stick the compass needle into the right point, span the compass pencil over to the left point, and now draw an arc that leads *upward* from the left point. . . . Now stick the compass needle into the *same place* on the *left point* where your arc begins, span your compass to the right point, and draw another arc leading upward from there. Do you see where the two arcs intersect at a new, third point? . . .

- To complete your already-existing figure, stick the compass needle in the *upper point* you just made. Now connect the two points you made at the beginning with an arc. . . .

- Now there is a figure on your cardboard. It has three *corners* and is bounded by three *arcs*. . . . Cut out your cardboard figure and place it in front of you on your desk. . . .

---

17   The *width of a planar shape* is the distance between any two distinct parallel lines, each having at least one point in common with the shape's boundary (but none with the shape's interior).

- To conclude, take two *strips of wood*. Place one wooden strip on the table so that it touches your figure from the *left*. Then place the other strip of wood on the table so that it touches your figure from the *right*. Align the wood strips parallel to each other and glue them to your desk. . . .

- Your figure is now clamped between the wooden strips but are you still able to *move* it? . . . Is it even possible for you to *rotate* it? . . .

- What did you imagine during this exercise in Mathematical Imagining?

This exercise deals with the simplest of this type of planar shape, the *Reuleaux triangle.* It is formed from the intersection of three circles of the same radius, each centered at a different corner of an equilateral triangle. At the end, students will cut out their figure and test the property of constant width.

Comments

The constructed Reuleaux triangle can actually be rotated, without getting stuck between the parallel wood strips or, for that matter, even touching one of them. However, the triangle's center (or more exactly said, its center of gravity) does move up and down and also back and forth.

The imagining task makes use of strips of wood, as students are most likely not familiar with tools like monkey wrenches or sliding calipers. The disadvantage of the wooden strips is that it is difficult to control their parallel position, even though they are glued to the table. This is why I recommend that, after finishing the exercise as presented here, have your students experiment with calipers so that they can really convince themselves of their figure's constant width.

That the Reuleaux triangle actually does have *constant width* is due to its construction. When a (real) Reuleaux triangle is clamped between a caliper's jaws and then continuously rotated, the following can be observed. When one corner of the Reuleaux triangle touches one of the caliper's jaws, most of the time, the other jaw will make contact with the opposite, arc-shaped side of the Reuleaux triangle. The exception is when the second point of contact is at one of the corners on the opposite side. In this case, not one but both jaws are touching corners. As the width remains invariable during the entire rotation process, the result is even when a plane figure has the same diameter in all directions; it is not necessarily a circle.[18]

---

18   According to Feynman (1989, 122), such a mistake (design flaw in the O-rings) might have been partly responsible for the 1986 space shuttle *Challenger* disaster.

A common occurrence in practice is that some students imagine the sides of the Reuleaux triangle as being too puffy or bulged. This could be connected to the fact that students were supposed to construct their Reuleaux triangle out of arcs. In doing so, the midpoints of the arcs near the triangle's sides seem to slip toward that side (a similar phenomenon appears in exercise R3). Other students imagine the Reuleaux triangle in its entirety as being too round, particularly after they have succeeded in rotating it. In order to positively answer the question in the instructions about whether they can rotate their figure, these students appear to "sand down" their figure to make it "behave," and this makes their figure more circular. How to cope with such counterproductive mental images in class is described in Chapter 2 in the context of another imagining task.

Mathematical follow-up questions

Once students have made first inquiries into the Reuleaux triangle, the imagining task makes possible a wide variety of other activities and questions.

- The imagined construction can be transferred to all planar regular polygons with an uneven number of corners. Students draw arcs centered on each vertex passing between the endpoints of the opposite side of the polygon. This leads to an infinite number of further Reuleaux polygons with constant width.

- Why isn't it possible to transfer the construction presented here to regular polygons with an even number of corners?

- A circle's circumference is $\pi \cdot d$. What is the circumference of the Reuleaux triangle? Surprisingly, the answer is also $\pi \cdot d$ and not just in this case: the circumference of all convex figures of constant width is always the same length, namely, $\pi \cdot d$ (*Barbier's theorem*).

- The area of a Reuleaux triangle is $\frac{1}{2} \cdot \left(\pi - \sqrt{3} \right) \cdot d^2$. This is the minimum area for a planar convex shape with a diameter of d (*Blaschke-Lebesgue theorem*). It is about 10 percent smaller than that of a circle, which, because of its isoperimetric property, realizes the maximum amount of area to the given diameter. This might be the reason, for example, that more economically produced buttons are often in Reuleaux triangle shape. Regardless of their orientation, they can fit through buttonholes, despite not being round (see Figure 5.8). Most likely similar considerations were behind the design and minting of the British twenty and fifty pence coins, which are shaped like a Reuleaux heptagon.

- Analogous to the task here, is it possible to construct from a tetrahedron a solid shape with constant width? For an exercise to do this, see exercise C11.

FIGURE 5.8 A button in the shape of an almost perfect Reuleaux triangle

Because of its particular characteristics, the Reuleaux triangle is often found in many technical devices and tools, such as film projectors, Wankel engines, belt tensioners, and monkey wrenches. A further practical use of this shape is drills that have Reuleaux triangle–shaped bit profiles capable of drilling holes that are almost square.[19]

---

19 To watch a silent animated film about this, see http://www.etudes.ru/en/etudes/reuleaux-triangle/ and http://www.etudes.ru/en/etudes/drilling-square-hole/. For more on drilling square holes, see the comments to imagining task PS7.

C8   DUALIZING A REGULAR TILING OF THE PLANE

DIFFICULTY LEVEL: ★ ★

MATHEMATICAL IDEA: A regular tiling with squares is invariant under dualization.

PREREQUISITES:
- Regular tiling with squares.
- Center of a square.

Duality and symmetry are both basic mathematical principles. Duality contributes connections not only between various objects, structures, and theorems to geometry but also to other mathematical disciplines as well, including graph theory, linear algebra, functional analysis, and logic. However, there doesn't seem to be a definition of duality that unites all various discipline-dependent perceptions of this principle.

This exercise in Mathematical Imagining introduces your students to duality through the simplest regular tiling in the plane so they can become familiar with the concept and try it out.

- Imagine you have a stack of large, *square* cards in front of you. Place such a card in front of you, in the *middle* of your desk, so that its sides are *vertically* and *horizontally* aligned to the edges of the top of your desk. . . .

- Align more cards around this one until the top of your desk is *completely* covered with squares with no gaps between them. . . . From your first square, there is a *vertical column* of squares that runs to the front and back of your desk and *another column* that runs across it horizontally left and right. . . .

- Focus on the places where the *edges* of neighboring squares *line up*. . . . You see a *net* of vertical and horizontal lines all over your desk. . . .

- Look once more at your *first* square, in the middle of your desk. Pick up a pen or pencil and mark the square's *center* with a dot. . . .

- Now *connect* this center with the four centers of the square cards that are *directly* above, below, left, and right of it. . . . In this way, you now see a *cross*. . . .

- Now do the same thing for *all* the other squares on your desk. Connect each square's center to the ones in the four neighboring squares. . . .

- What pattern do the lines you drew build? . . .

- What did you imagine during this exercise in Mathematical Imagining?

Comments

When the centers of adjacent squares of a regular square tiling are connected, the same tiling results again. This is why this tiling is also known as *self-dual.* The reason for this phenomenon is that, under dualization, the roles of vertices and faces are interchanged, and in regular tiling with square tiles, four tiles meet at every vertex. Students can verify this with graph paper.

Mathematical follow-up questions

There are two further regular, planar tilings. One consists of regular triangles, the other of regular hexagons (honeycomb pattern). Should one of these two tilings be dualized, it results in building the pattern of the other one, meaning the faces of one tiling correspond to the vertices of the other (see Figure 5.9). In other words, two consecutive dualizations of the tiling return to the original tiling. In a similar way, tilings can be dualized when, at every vertex, the same total number of regular polygons—not necessarily all the same kind—meet (*semiregular tilings*).

However, not only plane polygons and tilings can be dualized. If you want to dualize a spatial Platonic solid, every face with n sides corresponds to a vertex where n edges meet, and edges between pairs of vertices correspond to edges between pairs of faces of the other. If you dualize a cube, for instance, you get an octahedron. A tetrahedron though, is self-dual (ignoring the mirror-image orientation of the dual), as are, in fact, all pyramids with a polygonal base.

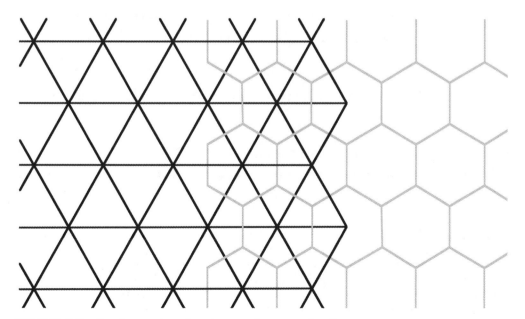

FIGURE 5.9 Dualizing the tiling by regular triangles/hexagons

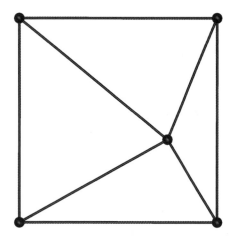

FIGURE 5.10 Schlegel diagram of a square pyramid

It is helpful to dualize convex polyhedrons using graph theory. Here, the edge structure is first translated into a graph: vertices are depicted as points, edges as line segments, and faces as bounded areas. This gives rise to a *Schlegel diagram* of the polyhedron (see Figure 5.10).

For the dualization, a new point is entered in each bounded area, and these new points that are in contiguous areas are then connected with a new edge. The dual polyhedron results when this second graph is back-translated into space. I describe how to work geometrically with the dualized polyhedron in the comment to imagining task C4.

The concept of duality as described here is closely connected to *projective geometry*. Thus, the following principle of duality applies in the projective plane. When in a true incidence relation statement about points and lines, the term *points* is exchanged with *lines*, *lies on* with *passes through*, and *intersect* with *connect*, the result is another true statement.[20]

Thus, a straight line in projective geometry is incident with an infinite number of points (a *point range*), and dual to this, a point is incident with an infinite number of lines (a *line pencil*). Curves, such as conic sections, can be seen as made from an infinite number of points or an infinite number of lines (the tangent lines to the curve). Two statements that are dual to each other are the theorems of Pascal and Brianchon:[21]

- *Pascal's theorem*: In a hexagon inscribed in a conic section, the three intersection points of pairs of opposite sides lie on a straight line (the *Pascal line*).

- *Brianchon's theorem*: In a hexagon circumscribed around a conic section, the three joining lines of pairs of opposite vertices meet in a single point (the *Brianchon point*).

Students can explore and deepen their understanding of this type of dualizing not only with graph theory but also through further exercises in Mathematical Imagining.

---

20 Parallel lines seen in perspective appear to intersect. Accordingly, planar projective geometry posits that each pair of parallel lines shares an *infinitely distant point*. All these infinitely distant points together form the *infinitely distant line* of the plane. While the parallel axiom no longer holds, every pair of lines has a unique intersection point. As every pair of points has a unique joining line, the axioms of plane projective geometry are fully symmetric with respect to point and line and with respect to joining and intersecting. Thus, all consequences of these axioms can be "dualized." For details about projective geometry, see Hilbert and Cohn-Vossen (1952, 112–132).

21 For more details, see Glaeser and Polthier (2020).

C9   AMBIGUOUS PYRAMID

DIFFICULTY LEVEL:	★ ★
MATHEMATICAL IDEA:	The ground plan of the wire-frame model of a square pyramid can be seen in different ways.[22]
PREREQUISITES:	• Ground plan (or view from above) of a square pyramid.

There is a very long tradition of using proofs without words to demonstrate mathematical truths, as shown by diagrams on ancient clay tablets that show the Pythagorean theorem.[23] Contemporary math lessons, too, are increasingly making use of illustrative sketches and self-explanatory graphics. However, such visuals do not automatically open the door to knowledge and understanding. Only students who have and can activate the key of relevant background knowledge can understand proofs without words. The use of sketches and other visual aids doesn't eliminate the need for mental effort, as such visuals are a form of thinking itself. In other words, visual aids are also content our students have to learn. Before students can give meaning to the illustrations, they must first understand what inherent expectations and knowledge the visuals contain and are trying to convey.

Psychological experiments with optical illusions designed to investigate how visual perception functions also point to the importance of necessary preknowledge. For example, the well-known ambiguous figure *My Wife and My Mother-in-Law* has been used in visual perception testing. Whether you see a young woman's profile looking away from you, or that of an old woman's looking toward you, depends on your perspective.

Perspective is also key in this imagining task that describes the "ground plan" of a square pyramid. Students can interpret this mental image as showing the pyramid from either an overhead or a bird's-eye view or from the worm's-eye perspective looking up from beneath the pyramid. This is because when a three-dimensional figure is drawn on, or projected in the plane, it is not clear where front and back are. In other words, the ambiguousness of this optical illusion is grounded in the projection.

---

22   The *ground plan* of a three-dimensional figure is the view from above, that is, an orthographic projection along the vertical direction.

23   The square of the hypotenuse or the squares of the other two sides can be complemented by four copies of the original triangle to two congruent squares (Nelsen 1993, 3).

- Imagine a *square* on a piece of paper. . . . Now sketch in the square's *diagonals*. Because the diagonals cross in the middle, there is an intersection point. . . .

- Let your square *come off* the drawing surface. Now instead of a square with its diagonals, you are looking at a *pyramid* from *above*: the former intersection of the diagonals is now the apex of your pyramid, whose edges are pointed toward you. . . .

- Now see your figure as if you are looking up from beneath the pyramid. Imagine that you can *see through* its base and watch how its edges leave the base, moving away from you to join together in the apex. . . .

- After you have looked at your pyramid from both of these perspectives, begin *switching back and forth* between them. Look at your pyramid *from above*, and then switch to look from *below*. . . .

- Try now to *keep hold of* the perspective you have *right* now; don't switch back and forth between the two views any more. . . .

- To finish, let your image become *flat* again, remove the midpoint and the diagonals, and let your square fade away. . . .

- At first, was it more challenging to view the pyramid from above or below? Did that change as you switched perspectives?

- What did you imagine during this exercise in Mathematical Imagining?

Comments

In a ground plan of a square pyramid, neither the meaning of the individual segments nor their relation to each other is given. This is why it is possible to interpret it in different ways and use it as a simple optical illusion.

Most of your students will easily be able to switch between the two perspectives right off the bat. Sometimes, they first imagine what a pyramid looks like from the side perspective, where it points up (or down), before they change back to viewing it from above and from below. Often, students' pyramids change proportions during the exercise. It gets a bit taller with each switch of perspective, and the summit becomes a little more pointed (compare to the counterproductive mental images in task C2). Once students start "switching," some of them, though, may have problems stopping, and the changing back and forth of perspective takes on a life of its own.

Only one time in my classes did this imagining task bring up negative associations. A student mutated the inside of the pyramid into a dark, bottomless pit that plunged down beneath her. Without even being aware of it, she opened her eyes in order to resist this mental image and the anxiety it was causing.

Mathematical follow-up questions

The depiction of a square with filled-in diagonals can also be used for additional interpretations other than "this is the ground plan of a square pyramid." For example, a regular octahedron has the same bird's-eye and worm's-eye views, and also its front and side views are those of squares with their diagonals. What other spatial interpretations are possible with this ground plan? (Exercise PS6 examines under which conditions a solid can be unambiguously reconstructed from its projection.)

Orthographic projections of other solids also allow for multiple spatial interpretations. The first confrontation with an oblique projection of a cube often presents students with a challenge: How should I interpret or "see" this (Figure 5.11, *left*)? Try to see in it a polyhedron with a hexagonal base (the figure's entire outline) and topped with a smaller parallel square surface (the inner square)! (The polyhedron's lateral faces consist of two triangles and four quadrilaterals.) For someone who is predisposed to "see" a cube, it can be difficult to see another possible shape

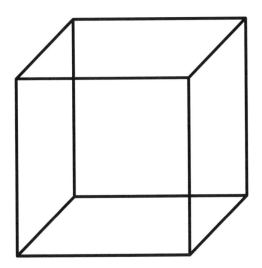 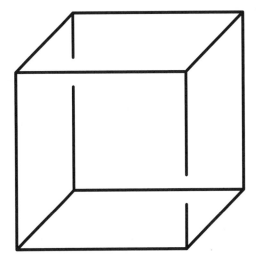

FIGURE 5.11   Necker cube (*left*) and Escher cube (*right*)

in it. It can be equally difficult for students to recognize a spatial cube in the oblique projection the first time they are confronted with it. The conventional interpretation of this oblique projection is far from being obvious and clear-cut; instead, it must be learned.

The optical illusion of the *Necker cube* (Figure 5.11, *left*) is also based on the fact that the oblique projection of a cube does not have a unique interpretation. Just like the pyramid projection above, the Necker cube doesn't contain any features to indicate depth, such as shadows or occlusion of the edges. This means that each of the two larger squares could be perceived as being the front or the back of a cube. Accordingly, under prolonged observation, the visual perception and interpretation of the Necker cube flips or switches. If at first you perceive a cube positioned below and to the left as seen from above and to the right, it can change quickly to become a cube positioned above and to the right and seen from below and to the left. It seems as if perception becomes unstable when confronted with this cube, as though our biological sight and our psychological perception have become uncoupled.

An artistic realization of the ambiguous interpretation of the oblique projection of the cube is from the Dutch graphic artist Maurits Cornelis Escher. His lithography *Belvedere* (1958) depicts, among other themes, a network of edges in which the edges are occluded in a way that suggests contradictory cube perspectives within one and the same image (Figure 5.11, *right*).

C10   CONSTRUCTING A DODECAHEDRON

DIFFICULTY LEVEL:	★ ★
MATHEMATICAL IDEA:	A dodecahedron can be constructed out of two congruent objects each with five attached flaps.
PREREQUISITES:	• Familiarity with the shape of a regular pentagon.
	• When five congruent pentagons are glued to the sides of a sixth identical pentagon, wedge-shaped gaps appear between the attached pentagons.

J ust like with an icosahedron (exercise C1), only a few experts can easily call up a mental image of a dodecahedron. This is especially so because dodecahedra are formed out of pentagons instead of triangles, which makes imagining this shape quite a challenge.

This imagining task also reads like an arts-and-crafts project for making a cardboard model of this solid. Students begin with a regular pentagon and then attach five other pentagons—one to each of the first one's five sides. Again, just like in the icosahedron construction, students then replicate the resulting figure and then put the two copies together to create their dodecahedron.

- Imagine a *regular pentagon* made out of cardboard and place it on your desk in front of you. . . . Now glue to *each one* of your pentagon's five sides another regular pentagon *that looks exactly like* your first one. . . . You have a figure that looks a little bit like a large, regular pentagon, except it has *wedge-shaped gaps* between the five pentagons you attached to the first one. . . .

- Look closely at one of those gaps. *Glue* the two edges together that form this gap. To do this, you need to slightly *fold* upward the two pentagons that are to be glued. . . . Now close the other four gaps in your figure in exactly the same way. Now you have a kind of *bowl* with five corners. . . .

- Now construct *another* bowl like this that looks exactly like your first one. When you are finished, *turn* this second bowl *over*. . . .

- To finish, put your turned-over bowl on top of your first one like a *lid*. This gives you a *closed* solid. . . .

- What does your solid look like? . . . How many edges come together at one vertex? . . . How many vertices does your solid have? . . .

- What did you imagine during this exercise in Mathematical Imagining?

Comments

As in "Constructing an icosahedron" (C1), the goal of the first question is to have students visualize the solid. Students can use their constructed bowls to see that, at each vertex, the same number of edges and faces come together. Students can also count the exact number of these elements. To count the vertices, it is helpful to imagine the dodecahedron as a combination of their constructed bowls, while taking into account that the vertices on the two bowls' rims now coincide and should, therefore, not be counted twice.

Students are not as familiar with a regular pentagon as they are with the triangle shape underlying an icosahedron. This means they may not accept the statement in the instructions that gluing two pentagons onto a third results in a gap (rather than a perfect fit or even an overlapping). This step often challenges students and hinders them from progressing in the following imagining steps. In that case, you might want to have students arrange three real cardboard pentagons to create one perceptible gap and then try imagining again.

Mathematical follow-up questions

For all Platonic solids, both Euler's polyhedron formula and the equation $q \cdot v = 2 \cdot e = p \cdot f$ apply (where q is the number of p-gons that meet at a vertex).[24] This means that, just by knowing the number of edges or faces per vertex ($q = 3$), and the number of vertices ($v = 20$), students can also determine the number of edges ($30 = e$) and all the other quantities of their dodecahedron.[25]

Broadly speaking, Platonic solids do not need three known quantities to be defined; two are sufficient. This is why one specifies either two of the three numbers for v, e, and f or, more commonly, the two numbers p and q. Accordingly, the *Schläfli symbol* $\{p, q\}$ is used to denote regular polyhedra (with p-gonal faces, q around each vertex).[26]

The construction procedure described delivers a model of a dodecahedron that is collapsible and, therefore, easy to transport. To create this model, have your students make the two notched figures out of cardboard as per the instructions, and make sure that the five pentagons hang flexibly on the central pentagon, but stop

---

24   A p-gon is a polygon with p sides.

25   According to *Cauchy's rigidity theorem*, convex polyhedra are determined in all their dimensions (distances from vertices, volumes) as soon as you know that its faces are congruent and how many faces meet at each vertex (see also footnote 2).

26   For the calculation of v, e, and f from the numbers p and q, see Coxeter (1973, 13–14).

before instructing students to lift their pentagons into a bowl shape. To build the dodecahedron, they can lay one of their flat figures on top of the other and then rotate them with respect to each other so that the vertices of one lie equally spaced between the vertices of the other. They can then stretch an elastic, alternating over and under the combined figure's vertices, so the rubber band is higher than a lower vertex and lower than a higher vertex. The band's tension will cause the figure to pop up in its three-dimensional form. Because of the fivefold symmetry of each of the two bowls, the elastic band forms a regular decagon.

As with the cube in exercise C2, students can also investigate cross sections of the dodecahedron. Can they slice their figure so that the cross section is a hexagon? Slicing the solid along a plane of mirror symmetry results in a hexagonal cross section. However, it is not regular, as two sides of it run along the dodecahedron's edges, and four sides run along the altitudes of the pentagonal faces. To obtain a regular hexagon, students must slice the dodecahedron perpendicular to its threefold rotation axis (which passes through two opposite vertices). This cross section also divides the dodecahedron into two congruent halves, just as the regular decagon does.[27]

Students can also obtain a square cross section when they cut their dodecahedron. For that, they imagine that they are looking at a wire-frame model of their figure along the twofold rotation axis running between the midpoints of two opposite dodecahedron edges. Focus on the closer of these two opposite edges. At its endpoints, this edge forks and turns into two pairs of new edges. The four endpoints of the new edges span a quadrilateral. This is actually a square, since the vertices lie in a plane, each of its edges is a diagonal of one of the pentagonal faces, and all its angles are equal (by mirror symmetry in the plane joining the original two opposite dodecahedron edges).[28]

As a consequence, students can inscribe cubes into the dodecahedron, whereby each face of the cube is covered with a hipped-roof-type structure. This can be done in various ways. The dodecahedron has thirty edges, so there are fifteen rotation axes. As three of these rotation axes are involved with each cube, five cubes can be inscribed in this way into the dodecahedron.

One other Platonic solid can also be inscribed into a dodecahedron. It arises when the dodecahedron $\{5, 3\}$ is dualized. Because, at each vertex, three regular pentagons meet, at every vertex of the dual solid $\{3, 5\}$, five triangles meet. As a

---

27   For illustrations of these two bisections, see Weisstein (2003, 813).

28   For cross sections of a dodecahedron, taken perpendicular to one of the axes of two-, three-, or fivefold symmetry, see Coxeter (1973, 237).

result, the icosahedron $\{3, 5\}$ is dual to a dodecahedron $\{5, 3\}$. (For more on the icosahedron, see C1, and for more on the dualization of solids, see the comments to C8.)

As already mentioned, the wedge-shaped gaps between two attached pentagons can also lead to further questions. The gaps have a vertex angle of $36°$, because the interior angles of a regular pentagon are $\frac{5-2}{5} \cdot 180° = 108°$. If the gaps become closed, the three pentagons meet at one point and create a bowl-shaped form. This piece of surface is topologically equivalent to a piece of a sphere.

If, instead, students try to arrange four regular pentagons around a point, then because the angle-sum at the point exceeds $360°$, the arrangement cannot remain in the plane of the desk and pushes into the surrounding space. The pentagons can no longer join at one point without warping, a behavior not seen with just three pentagons. The common edges of adjacent pentagons are no longer all pointing up or down; rather, they point alternately away from the plane of the desk and toward it (Figure 5.12, *left*). This property means that this piece of surface is kind of a *saddle surface*—a classic example of a surface with *negative Gaussian curvature*. It can be extended infinitely by embedding it in the *hyperbolic plane* (Figure 5.12, *right*). This regular tiling is denoted with $\{5, 4\}$ at each vertex, where four regular pentagons meet.[29]

In the Euclidean plane, there are only three regular tilings (see C8). In contrast, there are an infinite number of regular tilings of the hyperbolic plane. This means also that the hyperbolic plane allows tilings not only with four but also with five or more regular pentagons "crowding around" a vertex, resulting in a hyperbolic

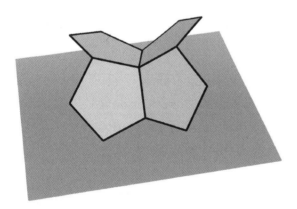

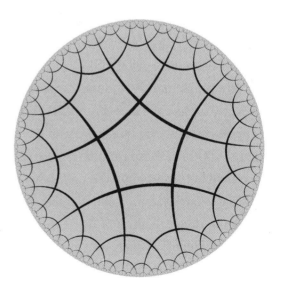

FIGURE 5.12 Regular tiling $\{5, 4\}$ *left* (above), partially shown in Euclidean space; *right*, shown in the hyperbolic plane using the Poincaré disk model

---

29   For details on the hyperbolic plane, see Weeks (2020, 135–148).

tiling in each case. The same applies for five and more squares around a vertex, which also then provides for an infinite number of regular tilings in the hyperbolic plane.[30]

This happens because only in hyperbolic tilings does the sum of the interior angles of the polygons that meet at a vertex exceed a full angle. If, at every point in a tiling, q regular p-sided faces meet and push into the surrounding space as the four pentagons above (see Figure 5.12, *left*), then the combined angle has to satisfy $q \cdot \left(\frac{p-2}{p} \right) \cdot 180° > 360°$. This leads to the inequality $\frac{1}{p} + \frac{1}{q} < \frac{1}{2}$, which with $p = 5$ is satisfied for $q = 4, 5, 6, \ldots$ However, it is also true for $\{p, q\} = \{3, 7\}, \{3, 8\}, \{3, 9\}, \ldots,$ of for $\{p, q\} = \{4, 5\}, \{4, 6\}, \{4, 7\}, \ldots$ and so on. In other words, there are an infinite number of other regular tilings in the hyperbolic plane besides the tiling $\{5, 4\}$.[31]

---

30 The program KaleidoTile (http://www.geometrygames.org/KaleidoTile/, for Windows and Mac OS X, Free-ware) lets you experiment with not only regular and semiregular tilings in the hyperbolic plane but also all regular and semiregular tilings of plane and spherical geometry. In the case of the sphere, you get Platonic and Archimedean solids.

31 Appropriate considerations result in $1/p + 1/q = 1/2$ for the Euclidean case and in $1/p + 1/q > 1/2$ for the spherical case. Consequently, there are only a finite number of plane tilings and only a finite number of plane polyhedrons that are topologically equivalent to spheres (Coxeter 1973, 5, 11; Hilbert and Cohn-Vossen 1952, 292–293).

C11   CONSTRUCTING A SOLID OF CONSTANT WIDTH

DIFFICULTY LEVEL: ★ ★

MATHEMATICAL IDEA: The sphere is not the only convex solid with constant width.

PREREQUISITES:
- Constructing a Reuleaux triangle (see C7).
- Rotating a regular triangle in space.

When you place heavy objects on loose rollers that have a circular cross section, you can transport the objects easily—and without shaking them. Roller conveyors are based on this same principle. Consistent with my comments to exercise C7, heavy objects could also be conveyed using rollers with a Reuleaux triangle–shaped cross section. If the rollers are on a level surface, the objects also move without vibrating because of the constant distance to the level surface. However, because of their cylindrical shape, such rollers can't roll back and forth in any arbitrary direction.

This leads to the question of what a genuine spatial analog of a Reuleaux triangle looks like. In this exercise, students will construct such a solid by rotating a Reuleaux triangle around its axis of reflection. The result is an object that is not a sphere but still has constant width.

- Imagine a Reuleaux triangle. It has three vertices and its edges bulge outward a little bit. Imagine that it is made out of cardboard and is lying in front of you on your desk. . . .

- The figure is *planar*. Now, you are going to use it to construct a *spatial solid*. To do this, first look at one of your Reuleaux triangle's *axes of reflection*. . . . This axis goes from one vertex through the midpoint of the opposite circular arc. . . .

- Now focus on these two points—one vertex and the opposite midpoint. Pick up your cardboard Reuleaux triangle by these two points and hold it in the air in front of you. . . . Give your cardboard triangle a spin so that it *rotates around* this axis of reflection. . . . Observe which points in space your Reuleaux triangle touches during the rotation.

- What *shape* does the resulting solid have? From which *parts* does it seem to be made of? . . .

- Place your solid on your desk and place a *wooden board* on top of it. Slide the board *back and forth* on the solid. Does the board wobble? . . .

- What did you imagine during this exercise in Mathematical Imagining?

Comments

The solid of revolution consists of a spherical cap and a piece of a spindle torus.[32] The spindle torus piece takes its start at a point on the rotation axis and ends where it meets the spherical cap in a circle. If this solid is clamped between two parallel plates, it will touch the plates either with its tip and the spherical cap or with its circular edge and the spindle torus. Because of the properties of the underlying Reuleaux triangle, this three-dimensional solid also has constant width.[33]

This astonishing property can be experienced this way. Put a sheet of glass on top of three such solids and use them to translate the glass back and forth. Because of the three support points, the glass remains stable on top of the solids, and because of the rollers' constant width, the glass moves completely parallel to the desk—vibration free—as if it were being rolled atop three spheres.

Mathematical follow-up questions

Based on your students' preknowledge and skills, there are numerous ways to expand this task with follow-up questions and activities.

- Besides the Reuleaux triangle, students can rotate other Reuleaux polygons around one of their mirror axes. Students can generate an infinite number of solids with constant width.

- The volume of the rotated Reuleaux triangles can be determined, for example, by integral calculus. If the width of the solid is d, its volume is $V = \left(\frac{2}{3} - \frac{\pi}{6} \right) \cdot \pi \cdot d^3$. If one considers—analogous to the two-dimensional case—all solids of constant width d with rotational symmetry, then this

---

32   The ring-shaped surface commonly known as a torus is a surface of revolution that results from revolving a circle around an axis in the plane of the circle that doesn't intersect the circle. Here, however, in the case of the revolving Reuleaux triangle, the axis intersects the circle; the resulting surface is known as a *spindle torus* (Weisstein 2003, 2836).

33   For a historical model of the rotated Reuleaux triangle made of plaster, see https://sim.mathematik.uni-halle .de/modellsammlung/models/Dg-002/index.html.

example has the smallest volume of all. This result was first proven in 1996 (see also the remarks about Meissner solids that follow).

- The surface area can also be calculated with integral calculus and is $S = \left(2 - \frac{\pi}{3}\right) \cdot \pi \cdot d^2$.

- If we consider both, the volume and surface area of the rotated Reuleaux triangle can be used to check the theorem of Blaschke. This theorem relates the surface S and the volume V of a convex solid with constant width d: $V = \frac{1}{2} \cdot d \cdot S - \frac{\pi}{3} \cdot d^3$. As a consequence, in three dimensions the "boundary area" of a solid with constant width depends on its volume—in contrast to the planar case where two shapes of constant width may have different areas but never different perimeters (see Barbier's theorem in C7).

- Another method of constructing a solid of constant width draws also on the construction of the Reuleaux triangle, but in a different way. Above each of the four faces of a regular tetrahedron, a surface of a sphere is constructed that is centered at the opposite vertex. However, the resulting intersection of the four spheres—a *Reuleaux tetrahedron*—does not (yet) have constant width. When the solid, though, rotates between two parallel supporting planes, it mostly touches them with one of its vertices and with the opposite spherical surface. For such pairs of points, the distance is, by the construction, constant. If, though, the two contact points lie on two opposite edges of the Reuleaux tetrahedron, the distance increases. By coordinate geometry, the width for the midpoints of two such edges can be computed to be $\sqrt{3} - \frac{\sqrt{2}}{2} \approx 1.02$ times greater than the sphere radius d.

  Because the width of the Reuleaux tetrahedron is not constant when the supporting planes touch two opposite edges of the solid, some edges have to be rounded off by the following procedure: First, remove the piece of the Reuleaux tetrahedron that is located between the planes of two adjacent tetrahedral faces and that contains a curved edge of the Reuleaux tetrahedron. Then, the resulting gap is to be replaced by a new piece of surface. This new piece of surface is generated by rotating one of the cut edges from the last step—it is a circular arc—around the corresponding ordinary tetrahedron edge until it "lands on" the second such cut edge. This new piece of surface, obtained by rotating the circular arc where the wedge of surface was removed, is a piece of a spindle torus.

FIGURE 5.13 Meissner solid with rounded edges surrounding a face

If you follow this procedure to round off each of the three edges meeting at a vertex, you will obtain the first type of *Meissner solid*. When, instead, you round off three edges surrounding one of the faces (see Figure 5.13) you obtain a different, noncongruent *Meissner solid.* Both solids have constant width and touch two parallel planes either with a vertex and the antipodal spherical surface or with a rounded and a nonrounded edge of the solid.

The volume of the two Meissner solids is $\left(\frac{2}{3} - \frac{\sqrt{3}}{4} \cdot \arccos \frac{1}{3} \right) \cdot \pi \cdot d^3$, and using Blaschke's theorem, their surface area is $\left(2 - \frac{\sqrt{3}}{2} \cdot \arccos \frac{1}{3} \right) \cdot \pi \cdot d^2$.

Thus, the Meissner solid not only has a smaller volume than the rotated Reuleaux triangle but is even conjectured to have the smallest volume of all solids with constant width d. This claim has yet to be proven.[34]

---

34 For photos of both Meissner solids, see https://sim.mathematik.uni-halle.de/modellsammlung/models /Dg-003/index.html or Hilbert and Cohn-Vossen (1952, 215–216). For more background information, see Kawohl and Weber (2011) or https://www.swisseduc.ch/mathematik/geometrie/gleichdick/index-en.html.

C12   CONSTRUCTING THE CANTOR SET

DIFFICULTY LEVEL:	★ ★ ★
MATHEMATICAL IDEA:	Certain geometric objects are intermediate between a point and a line segment.
PREREQUISITES:	• Dividing a line segment and removing a subsegment.
	• Continuously repeating an action.

Our everyday experience supports the assumption that removing something from an object makes the object smaller. Thus, it appears impossible that a geometric object could contain the same number of points as a line segment but have a length of zero. The Cantor set shows, on the contrary, that this is possible.

In this exercise, students remove the middle third of a line segment and, then, in turn, remove the middle third of each of the remaining segments. When this process is repeated an infinite number of times, what remains at the end is called the *Cantor set.*

- Imagine that you are drawing a horizontal line segment that is about as long as your forearm is wide. . . .

- *Divide* your line segment into three sections that all have the same length—and then *delete* the section that is in the middle. . . . Now you have *two* sections left over: the first one and the third one. . . .

- Now *divide* each of your two remaining sections into three equally long sections. And, once again, *delete* the section in the middle. . . .

- Now repeat this process of dividing into three sections and removing the middle third. . . . Keep doing this *over* and *over* and over again. . . .

- What does the object you created in this way look like at the end? . . .

- What did you imagine during this exercise in Mathematical Imagining?

Comments

If all boundary points of each "inner" section are retained (meaning only *open intervals* are eliminated), the result is a peculiar fragmented structure known as the *Cantor set.* It looks a bit like a very fine-toothed comb, which is why it is sometimes also called *Cantor comb.*

Because, in the process, students have to see increasingly smaller sections, they tend to change the proportions of the objects during the imagining task. For example, their mental image of the entire line segment keeps getting longer each time they move to the next, finer segmentation—with some students, exactly by a factor of 3. Another issue is that, sometimes because of the repetition involved in the mental actions (dividing one or more line segments and removing some of these), a resulting mental image of the action looks similar to the one that preceded it. This corresponds to the similarity of the whole Cantor set to its parts.

Mathematical follow-up questions

What exactly remains when the process of deleting continues infinitely? What is certain is that no single connected piece of the line segment, no matter how small, is contained in the Cantor set. Although the repeated division-into-three and deletion process does leave new line segments at each step, it does so only to target them in the next step. Therefore, sooner or later the process reaches every segment contained in the starting interval, and none of them "survives" intact.

If the original line segment is the unit interval $[0,1] = \{x \mid 0 \leq x \leq 1\}$, then it's possible to immediately identify some points that make it through the infinitely repeated process unscathed. For example, the points 0 and 1 survive not only the first instance of deletion but also all those that follow. In each step of deleting, only the points that are between the remaining intervals disappear. Similarly, the infinitely many points that become boundary points during the deleting process, such as $\frac{1}{3}$ and $\frac{2}{3}$, and also $\frac{1}{9}, \frac{2}{9}, \frac{7}{9}$, and $\frac{8}{9}$, and so forth, will "survive" and are thus contained in the Cantor set.

The Cantor set, though, consists primarily of quite different points. In order for students to get a sense of these, represent the elements of the Cantor set in base three. The numbers between 0 and 1 can be represented not only as decimal fractions but also in the form $\sum_{k=1}^{\infty} \frac{a_k}{3^k}$ (with $a_k \in \{0, 1, 2\}$). This is the *ternary* (or base three) *representation* of a number t, usually indicated with an index 3, for example, $\left(\frac{7}{9}\right)_{10} = 0.21_3$. Now, by eliminating all numbers that contain a "1" in their ternary representation (this means any number with $a_k = 1$ for one or more k's), we get an arithmetic description of the Cantor set. This elimination corresponds to deleting the middle third of the line segment. Does that mean boundary points such as $\frac{1}{3} = 0.1_3$ are not in the Cantor set after all? Such doubt is unfounded, because just as with terminating decimal expansions (where, for example, $0.7 = 0.6\overline{9}$), terminating ternary expansions can also be brought into an equivalent nonterminating form: $\frac{1}{3} = 0.1_3 = 0.0222..._3 = 0.0\overline{2}_3$. Consequently, although the number 0.1_3 is eliminated, the number $0.0\overline{2}_3$ will survive the process of elimination and with it the corresponding boundary point.

With the help of the base three representation, you can also specify elements from the Cantor set that are not boundary points, for example, $0.0\overline{2}_3$. This number corresponds to $\frac{0}{3} + \frac{2}{9} + \frac{0}{29} + ... = \frac{1}{4}$. The first 0 means that $\frac{1}{4}$ is in the lower third of the first subdivision; the first 2 means that it is in the upper third of the subdivision of that subsegment. Each subsequent 0 or 2 means that the point is contained in the lower or upper third of the current subdivision and hence is never in a (deleted) middle third. Thus, it is contained in the Cantor set.

How "big" is the Cantor set? If you calculate the length of all deleted intervals $\frac{1}{3} + 2 \cdot \frac{1}{9} + 4 \cdot \frac{1}{27} + ...$, through use of geometric series, you get

$$\frac{1}{3} + 2 \cdot \frac{1}{9} + 4 \cdot \frac{1}{27} + ... = \frac{1}{3} \cdot \sum_{n=0}^{\infty}\left(\frac{2}{3}\right)^n = \frac{1}{3} \cdot \left(\frac{1}{1-\frac{2}{3}}\right) = 1$$

(or, with the ternary expansion,

$$\frac{1}{3} + 2 \cdot \frac{1}{9} + 4 \cdot \frac{1}{27} + ... = 0.222..._3 = 0.\overline{2}_3 = 1).$$

Consequently, the length of the remaining Cantor set is 0—similar to that of a point. On the other hand, just as a line segment, the Cantor set contains an (uncountably) infinite number of elements. This is hardly imaginable—and not just for our students.[35]

As previously mentioned, each magnification of the Cantor set by the factor 3 reveals its original structure again. This property of self-similarity also works in reverse and can be used to generate the set, for example, on a computer. Therefore, the two contraction mappings $w_1 : t \mapsto \frac{t}{3}$ and $w_2 : t \mapsto \frac{2}{3} + \frac{t}{3}$ are applied to the unit interval $[0, 1]$. They shrink the unit interval by a factor of 3 and map it onto the two subintervals $\left[0, \frac{1}{3}\right]$ and $\left[\frac{2}{3}, 1\right]$. When the two contractions w_1 and w_2 are then applied to the union of the two intervals, four intervals are generated. In the limit, the Cantor set emerges once more. In a certain sense, the contractions w_1 and w_2 are like a photocopier set in "make multiple scaled-down copies" mode.

With self-similarity, we can broaden the general conception of dimension and define a *self-similarity dimension*. If a set consists of N scaled-down copies of itself and the shrinking factor for all copies is p, then the equation $p^d = N$ or $d = \frac{\log N}{\log p}$ holds for geometrical figures in Euclidean geometry (with dimension d; see Figure 5.14).

---

35   For the difference between geometrical size and set-theoretical cardinality, see exercise P4.

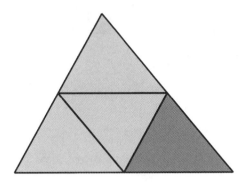
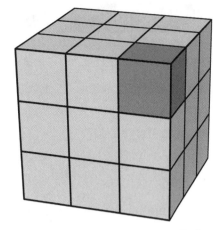

FIGURE 5.14 A self-similar object in the plane, and another one in space (*left* with $N = 4$ and $p = 2$, *right* with $N = 27$ and $p = 3$)

By construction, the Cantor set consists of two copies of itself ($N = 2$), each copy shrunk to one-third its original size ($p = 3$). Hence, you obtain a noninteger self-similarity dimension, which is $\frac{\log 2}{\log 3} \approx 0.63$. From this viewpoint, the Cantor set is also something in between a point ($d = 0$) and a line segment ($d = 1$). In other words, this is most likely the first *fractal* analyzed in mathematics.

Such fractals can also be constructed in a completely analogous way from objects that originate in a higher dimension. The *Sierpinski triangle* is about as simple to construct as the Cantor set. You can also have your students construct it by having them begin imagining an equilateral triangle and delete the triangle formed by the midpoints of its sides. When they keep repeating, ad infinitum, this divide and delete process on the triangles that remain after each step, a Sierpinski triangle results (see Figure 5.14, *left*). Its area converges to zero and its perimeter grows infinitely large. Here, the fractal dimension of the Sierpinski triangle is noninteger again but greater than 1. To be exact, it is $\frac{\log 3}{\log 2} \approx 1.58$.

An astonishing result arises when the same process is applied to a tetrahedron. As we have seen in a previous exercise, when the corners of a tetrahedron are cut off "just right," an octahedron results (the cuts pass through the midpoints of the tetrahedron's edges; see Figure 5.4, *right*). When instead the octahedron is "deleted" from the tetrahedron, four similar tetrahedra remain in which each has half the edge length of the original one. In the limit, a fractal results that is called the *Sierpinski tetrahedron*. Because $N = 4$ and $p = 2$, this fractal has a self-similarity dimension of $\frac{\log 4}{\log 2} = 2$. Contrary to the assumption that every fractal must have a noninteger fractional dimension, the Sierpinski tetrahedron is an example of a fractal with an integer self-similarity dimension.

C13   CONSTRUCTING A FOUR-DIMENSIONAL HYPERCUBE

DIFFICULTY LEVEL: ★ ★ ★

MATHEMATICAL IDEA: If you find just the right translation, you can build a
four-dimensional hypercube.

PREREQUISITES:
- Line, square, and cube.
- If a luminous point is moved straight ahead, the
trace of the light forms a line segment.

School mathematics reaches up to the third dimension. The fact that other, higher geometrical dimensions are thinkable rarely finds its way into school math lessons. At most, students may ask, "Is it true that the fourth dimension is time?" Adding time to the three spatial dimensions actually does create a four-dimensional space called *spacetime* in the theory of relativity. However, in it, time is singled out to play a "starring" role in comparison to the three geometrical dimensions. For this reason, spacetime cannot serve as a concept to help students experience the fourth dimension in Euclidean space, as in it, all dimensions are equal. And when we do bring in a fourth geometric dimension, how are we—and our students—to imagine a four-dimensional object? How can it be visualized? Does the fourth dimension "really" exist?

This exercise makes use of extrapolation to venture into higher dimensions, or, to be more exact, it uses sliding. The goal is not to have students visualize the four-dimensional ambient space but, rather, a four-dimensional object in this space. Students begin at a zero-dimensional point and, from there, slide the point to generate a line, a square, a cube, and finally a hypercube. To help students imagine this more easily, the point glows, and when they slide it straight ahead in an "unoccupied" spatial direction, it leaves behind a glowing line segment along its path. This luminous line segment is then slid in a farther, unoccupied spatial direction, where it leaves behind a path of light in the shape of a square and so on.

- Imagine a *point*. Your point is *glowing* brightly inside a dark room. . . .

- The light from your point has a special property: wherever your point moves, it leaves behind a path of light that *never, ever* goes out.

- Now as a first step, slide your point *straight* to the *right* for a distance that is about a *handbreadth* long. The trace of the light point is a *line segment* that is as long as your hand is wide. . . .

- The next step is to slide this glowing line segment a *handbreadth away* from you. The trace of the light is now a *square*, and its surface is glowing. . . .

- Now slide your glowing square a handbreadth in a *farther* spatial direction that you have not used yet: vertically *upward*. The trace of the light now forms a *cube*. Do you see how your cube is glowing as a whole solid? . . .

- To finish, try carefully to slide your cube into another spatial direction that you have not used before. . . . The trace of the light is *another* glowing object. . . . Can you see it?

- The glowing line segment had two endpoints, didn't it? When you slid it to make a square, the number of endpoints doubled, and you got an object with four vertices. So how many vertices does your cube therefore have? . . . And how many vertices does the final glowing object presumably have? . . .

- What did you imagine during this exercise in Mathematical Imagining?

Comments

Although it is quite difficult for students to achieve a stable mental image of the four-dimensional hypercube, the extrapolation-based construction in this exercise makes it possible for them to gain a first impression of this object, which is also known as a *tesseract.* Even if a student is not able to construct a clear mental image of it during the imagining task, they may still guess that the tesseract has sixteen vertices, as the number of these doubles with each sliding step.

Mathematical follow-up questions

This imagining task leads to a world of follow-up questions: How many and what kinds of objects form the boundary of a hypercube? How can it be drawn or otherwise depicted? Is it possible to generalize the Euler polyhedron formula so that it can be applied to this polytope, too?[36]

Based on the described sliding movements, the questions about type and number of boundary objects can be answered immediately. The four-dimensional hypercube is bounded by eight cubes. One cube is given at the start, and one results from sliding it. The tesseract's sixteen vertices lie on these two cubes. The remaining six boundary cubes are generated by sliding the six faces of the original cube.

---

36   The umbrella term *polytope* generalizes the terms polygon and polyhedron in higher dimensions (Coxeter 1973, 118, 126–127). A four-dimensional polytope is sometimes also called a *polychoron.*

At the same time, there are twenty-four squares belonging to the tesseract. These are generated through the last slide, whereby the beginning and ending cube each contribute six faces and the sliding of each of the twelve cube edges generates twelve new squares for a total of twenty-four. In this way, too, the hypercube has thirty-two edges, twelve coming from each of the beginning and ending cube, and eight new edges generated by sliding the eight vertices of the cube.

Thus, when you add together the number of vertices, edges, faces, (three-dimensional) solids, and four-dimensional objects in the tesseract, the result is $16 + 32 + 24 + 8 + 1 = 3^4$ zero- to four-dimensional components involved in the four-dimensional hypercube. In the same way, a point consists of 3^0 elements, a line of 3^1, a square of 3^2, and a cube of 3^3 elements. This observation that the number of elements triples for each step up in dimension applies in general. If an object that is to be slid has n elements, then through sliding (at the end) it will have n new additional copied elements, meaning all in all, the number of elements has doubled. Further, each of the n elements generates during the process a new element, which is one dimension higher than the generating element. This produces n further elements. In other words, the number of components triples per dimension—also in dimensions higher than four.

But you can also investigate the individual numbers of the component (vertices, edges, faces, and volume, etc.). By combinatorial considerations, an n-dimensional hypercube consists of $\binom{n}{k} \cdot 2^{n-k}$ k-dimensional components $(0 \leq k \leq n)$. Consequently, the desired number can be read on the coefficients of the expanded polynomial $(2 + x)^n$. Using the binomial theorem $\sum_{k=0}^{n} \binom{n}{k} \cdot x^{n-k} y^k = (x + y)^n$, students can deduce once again that an n-dimensional hypercube has a total of $\sum_{k=0}^{n} \binom{n}{k} \cdot 2^{n-k} = 3^n$ components.[37]

Euler's formula applies to three-dimensional polyhedra (see exercises C1 and C10). If students experimentally form the alternating sum of the number of vertices, edges, faces, and solids of the four-dimensional hypercube, they obtain $16 - 32 + 24 - 8 = 0$. The value zero not only results for the special case of the hypercube but also for all four-dimensional polytopes and in particular for the four-dimensional hypersphere (see P2). The reason for this is that the *generalized Euler polyhedron formula* $N_0 - N_1 + N_2 - N_3 + ... + (-1)^{n-1} N_{n-1} = 1 - (-1)^n$ applies (where N_k indicates the number of the polytope's k-dimensional boundary elements).[38]

---

37   See Coxeter (1973, 122, 292, 294).

38   For a generalization of Euler's polyhedron formula for objects with holes, see exercise PS7.

As with an ordinary cube, students can also unroll or unfold the four-dimensional hypercube (see Figure 5.15). To obtain its three-dimensional net, the cube must first be cut open on several of its twenty-four boundary squares. For the net on the right in Figure 5.15, seven pairs of square faces are still glued together (and, hence, cannot be seen), while the other seventeen square pairs appear separated (they form the boundary of the three-dimensional shape). They must be glued together when putting the hypercube back together. (One such pair is shown in dark gray in Figure 5.15, *right*.)

Salvador Dalí interpreted the tesseract's net artistically as a cross in his 1954 oil-on-canvas painting *Crucifixion (Corpus Hypercubus).* Which leads us to the question: How many nets are possible? While there are eleven nets of the three-dimensional cube altogether, in how many different nets can we "unfold" the four-dimensional hypercube?[39]

To draw the tesseract in central perspective (Figure 5.16, *right*), it doesn't have to be unfolded. Here, just like in the Schlegel diagram of an ordinary cube (Figure 5.16, *left*), you have to keep in mind that one of the eight boundary cubes is represented as the exterior of the figure shown in the diagram and, hence, is not shaded.

Although this illustration shows only the combinatorial properties of the tesseract, nevertheless, it allows for its realization and architectural implementation.

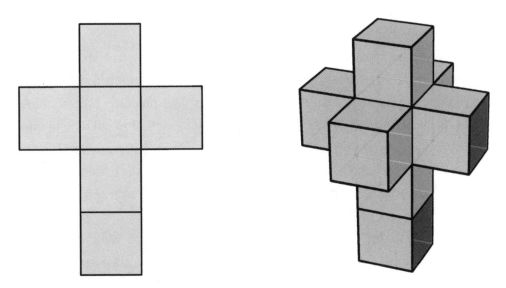

FIGURE 5.15 Net of the three-dimensional cube (*left*) and the four-dimensional hypercube (*right*)

---

39   There are 261 different nets of a four-dimensional hypercube (Weisstein 2003, pp. 2963–2964).

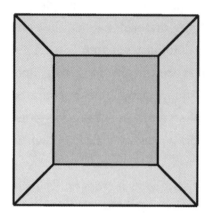 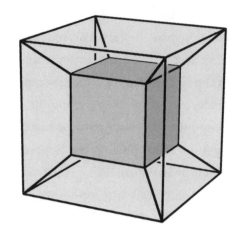

FIGURE 5.16 Central projection of the three-dimensional cube and the four-dimensional hypercube

(For example, *La Grande Arche de la Défense* (1989), a modern triumphal arch on the outskirts of Paris, is based on the central projection of a huge tesseract in three-dimensional space.

From the perspective drawing (Figure 5.16), you can infer, among other facts, that three cubes meet along each tesseract edge. This property is characteristic, as it may be used to define the tesseract. Begin with regular quadrilaterals and join three of these at each vertex to obtain a cube (with Schläfli symbol $\{4, 3\}$). If you then join three of these cubes together around each edge, you obtain a tesseract (with Schläfli symbol $\{4, 3, 3\}$).

Once students see the tesseract's combinatorial-topological structure with their mind's eye, they can work with and manipulate this polytope in various ways. They can, for example, explore the question of its dual polytope (see also exercise C8). Because in four-dimensional space, points are dual to solids and edges to faces, a polytope results that is constructed out of sixteen tetrahedra (one per each tesseract vertex) and has eight vertices (one per each of the tesseract's boundary cubes). This object is also often referred to as a *four-dimensional hyperpyramid* (with Schläfli symbol $\{3, 4, 4\}$).[40]

Again, students can investigate how to slice the tesseract. What solids result when we use a three-dimensional hyperplane to cut the tesseract perpendicular to

40   For the dual of the tesseract, see Coxeter (1973, 121–122, 148–149, 243). In four dimensions, there are a total of six regular polytopes (Coxeter 1973, 136–137, 292–293; Hilbert and Cohn-Vossen 1952, 144). You can watch an animation of them in a film by E. Ghys, "Chapters 4 and 5: The Fourth Dimension," Dimensions, http://www.dimensions-math.org/Dim_CH3_E.htm.

its longest diagonal (compare with exercise C2)? When we start from one vertex, a small tetrahedron emerges that increases in size until it morphs into a truncated tetrahedron. It keeps on growing until the moment the hyperplane goes through the tesseract's center point when the cross section becomes an octahedron. If students keep slicing "bit by bit," the octahedron morphs back to a truncated tetrahedron and finally ends as a vanishing small tetrahedron at the vertex opposite to the one where the slicing began. In other words, the tesseract can be cut in half so then a polytope section in the shape of an octahedron results (Figure 5.17, *right*).[41] Referring to Edwin A. Abbott's 1884 novella *Flatland: A Romance of Many Dimensions*, the described sequence of slicing solids is exactly what we would see if a four-dimensional hypercube would "dive vertex first" into our three-dimensional world.

There are many formulas (particularly in coordinate geometry) that students can directly transfer to higher-dimensional objects, for example, to determine length or angle sizes in the tesseract. For example, the distance between a pair of tesseract vertices that are the farthest apart from each other is, by using the generalized Pythagorean theorem, $\sqrt{1^2 + 1^2 + 1^2 + 1^2} = 2$ (see comment to R10). When carrying out such calculations, students should call to mind the corresponding mental pictures because merely performing an algebraic manipulation of symbols does not lead to a substantive comprehension of the four-dimensional hypercube.

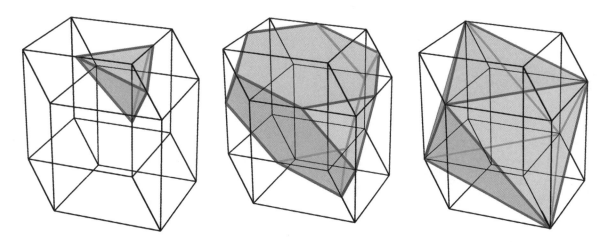

FIGURE 5.17 Slicing sequence of the four-dimensional hypercube (parallel projection)

---

41   For animated visualizations of the solids that arise by slicing a tesseract, see D. P. Cervone, "Some Notes on the Fourth Dimension: Hypercube Slices Viewed Corner First and Interactive Models of 4-Dimensional Objects," 4th Dimension, http://www.math.union.edu/~dpvc/math/4D.

Whether or not a fourth spatial dimension exists is not a question that can be or should be answered here. This exercise also does not assume that a higher dimension exists in the same physical sense as our three-dimensional universe does. Rather, the point here is to demonstrate that higher dimensions are conceivable through the interplay between imagination and abstraction and that we can draw conclusions about them.

Answering the question about whether something exists always depends on how deeply we grapple with and immerse ourselves in the subject. When we engage with a matter, we are in a sense investing it with "life" and it becomes part of reality. In this respect, the approach sketched here is perfect for giving students a living experience of the fourth dimension. In addition to imagining tasks, animated images and interactive animations of the hypercube can be helpful.[42] The value of "self-driving" a tesseract and exploring it is not to be underestimated, because doing it themselves serves to support students' development of a four-dimensional spatial vision.

---

42 A software suited to training four-dimensional vision is HyperSolids (http://pages.uoregon.edu/koch /hypersolids/hypersolids.html, for Mac OS X, Freeware). It depicts not only the tesseract wire-frame model but also those of all other convex, regular polytopes in four-dimensional space.

C H A P T E R

6

Problem-Solving Exercises in Mathematical Imagining

This chapter focuses on problem-solving exercises in Mathematical Imagining. The exercise instructions serve to embed a mathematical problem into a non-mathematical context and not primarily—as in the construction exercises of Chapter 5—to help your students construct a mathematical object. The questions asked are therefore open ones. These imagining tasks aim particularly to encourage mental experimenting and guessing.

Table 6.1 gives you an overview of all eight example tasks arranged according to difficulty level (more stars denote more difficulty). The notes and comments sections give you some background information about the tasks. The notes on tasks PS1 and PS2 discuss in detail the intended mental images and others that are likely to appear in practice. Each exercise concludes with possible follow-up questions and activities. (For the different types of imagining tasks and more on how the comments are structured, see Chapter 4.)

TABLE 6.1   List of problem-solving exercises in Mathematical Imagining

Exercise in Mathematical Imagining	Mathematical idea	Prerequisites	Pages
Constructing the Collatz sequence ★ ★	The Collatz sequence appears to lead sooner or later to a cycle.	• Mental calculation up to 160.	PS1 p. 126
Sliding a ladder ★ ★ ★	The path of the midpoint of a ladder sliding down a wall is a fourth of a circle.	• Handling a ladder (setting it up, causing it to slide down a wall). • Midpoint of a line segment.	PS2 p. 132
Sliding a drafting triangle ★	Suppose a drafting triangle is sitting on two nails so that its right-angle points downward. Then this right-angle corner moves along a semicircle as the triangle slides along the nails.	• Drafting triangle as an isosceles right triangle.	PS3 p. 139
Laying out a chain of squares ★ ★	Repeated actions can lead to different, yet equivalent, algebraic expressions.	• Arrange matches in a chain of squares. • Recognize individual squares within a chain of squares. • Mental calculation up to 50.	PS4 p. 142
Constructing the sequence of square number differences ★	The differences between consecutive square numbers form the sequence of odd numbers.	• Square numbers and odd numbers • Mental calculation up to ~200.	PS5 p. 145
Projecting and reconstructing solids ★ ★	Constructing a solid that casts one circular, one square, and one triangular shadow in the three coordinate directions.	• The shadow of a sphere in orthographic projection is circular, regardless of the direction of projection. • The orthographic projection of a right circular cylinder can have a shadow that is rectangular or circular.	PS6 p. 148

(continued)

TABLE 6.1   List of problem-solving exercises in Mathematical Imagining *(continued)*

Exercise in Mathematical Imagining	Mathematical idea	Prerequisites	Pages
Drilling through a cube ★ ★ ★	Three tunnels with square cross sections that cross each other pairwise at right angles intersect in a cube.	• Drill tunnels with square cross sections. • Retaining your orientation in up to three intersecting tunnels. • Mental calculations up to 40.	PS7 p. 153
Curving a straight line ★ ★	The straight line as the locus of all points lying equidistant to two fixed points becomes a hyperbola when one of the fixed points is increased into a small circle.	• Points that are equidistant to two fixed points form a straight line. • Increasing a point's diameter above zero produces a growing circle.	PS8 p. 158

PS1 CONSTRUCTING THE COLLATZ SEQUENCE

DIFFICULTY LEVEL: ★ ★

MATHEMATICAL IDEA: The Collatz sequence appears to lead sooner or later to a cycle.

PREREQUISITES: • Mental calculation up to 160.

Proficiency in mathematical analysis assumes an intuitive understanding of various phenomena concerning number sequences. In general, though, the study of number sequences is restricted to their convergence properties as needed in infinitesimal calculus. This narrowness is unfortunate because it neglects the possibility that particular number sequences might be of interest despite being divergent, for example, because they sooner or later become cyclical.

In this exercise in Mathematical Imagining, students will explore a number sequence ascribed to the German mathematician Lothar Collatz. The Collatz sequence is constructed as follows: Start with any natural number. Then, the next number is obtained by the following generating rule: If your number is even, divide it by 2; otherwise, if it is odd, multiply it by 3 and add 1. When you iterate this process, a sequence of numbers is generated.

During their exploration of the resulting Collatz sequence, learners meet the phenomenon of *cyclic* sequences. For every start number n they select, the sequence of resulting numbers appears always to lead eventually to the same cycle: $\ldots \to 1 \to 4 \to 2 \to 1 \to \ldots$. It has yet to be proven whether this is true for *all* possible starting numbers.

- Imagine any whole number *between two and twenty-four* and imagine writing it down so you can see it with your inner eye. . . .

- Now proceed with your number like this: If your number is *even,* divide it by two. If your number is *odd,* multiply it by three, and then add one. . . .

- Write down the resulting number so you can see it with your inner eye. Now, proceed in the same way as you did with the *original* number you chose: If your number is *even,* divide it by two. If your number is *odd,* multiply it by three, and then add one. . . .

- Keep repeating this procedure.

- What do you observe? . . . If you like, you can choose a new starting number and begin again.

- What did you imagine during this exercise in Mathematical Imagining?

Notes on intended mental images

You instruct your students to choose a whole number between 2 and 24 and apply the Collatz generating rule to it again and again. The intended mathematical question is to be found in the instruction to keep repeating the procedure and looking for a repeated pattern. This process produces a sequence of numbers. If this same operation is repeated on a few more starting numbers, your students will quickly guess that the sequence for any arbitrary starting number results in the same cycle $\ldots \rightarrow 1 \rightarrow 4 \rightarrow 2 \rightarrow 1 \rightarrow \ldots$.

There is a practical reason that the starting number should not be greater than 24. For starting numbers between 2 and 24, your students will arrive at the result number 1 after a maximum of 20 calculation steps (this happens for starting numbers 18 and 19.) Starting with 25 results in a sequence with 23 steps, while starting with 27 results in 111 steps! The numbers that appear in the Collatz sequences (for starting numbers 2 to 24) range from 1 to 160 (which is reached by starting numbers 15 and 23).

Unlike most of the other imagining tasks, the mathematical idea in this task is not really embedded in a geometrical or nonmathematical context of images and actions but, instead, is presented in an arithmetical, algebraic form. That's why the instructed imagined actions—mental calculation and writing down—are not particularly "real" activities (as, for example, moving or taping together imagined objects).

In order for your students to be able to observe and compare the elements of the Collatz sequence they obtained, they need to form mental images of the numerical values. These mental images serve the students here as support for their thinking. Instead of having students imagine writing down the numbers, you could also instruct them to imagine saying the numbers out loud. Depending on your class, you might be able to leave out this part of the instructions altogether and leave it up to your students to decide how they organize their mental images.

Notes on productive and counterproductive mental images that learners construct

Because of this imagining task's almost complete lack of a nonmathematical image and action context, students do not always find it easy to construct their own mental images. Nevertheless, they can develop mental images that are productive ones for constructing the intended mental images and answering the question.

- Students often imagine the set of real numbers spatially or geometrically. Accordingly, some students position the numbers they calculate on their imagined number line and jump back and forth between their numbers (recall imagined number line variations from Figure 1.1).

- A further group of mental images students construct connects to the cyclical nature of the sequence. Students experience this phenomenon quite strongly through the mental action of jumping. Some students construct a mental image of a number line that is somehow curved and then go on to modify the curve into a closed number circle. After they have traversed their circle line a few times, their images prior to the one of the constructed circle fade, and their mental image of the circle containing the three positions 4, 2, and 1 stabilizes.

- Some of the mental images students construct have to do with remembering the calculated numbers. In particular, during this imagining task, my students reported experiencing synesthetic color perception that they "saw" in connection to the calculated numbers and used the colors as a memory aid. Often, positive emotions accompanied the colors associated with the numbers.

- When students "see" an even number as the doubled amount of another number (e.g., 26 as $2 \cdot 13$), then they have already divided the number into two equal halves just through checking its evenness, thus saving themselves a calculation step. Similarly, after students multiply an (odd) number by 3 and add 1, the result is always an even number. That's why some students immediately divide it by 2 without checking it, and thereby save another calculation step.

For this exercise in Mathematical Imagining, there are not only counterproductive mental images deserving of mention but also difficulties with the imagining itself.

- When students have negative associations to the colors they "see" on the numbers 1, 4, and 2, it is hard for them to concentrate and remember the numbers.

- Some students start blithely calculating away. They linger at each calculated number only for as long as it takes them to make the next calculation step, erasing their prior work as they go. Therefore, they are unable to reproduce the number sequence afterward. It seems as if their calculation process is running automatically, meaning these students are not constructing the corresponding mental images and actions.

- Even when they have done the calculations correctly, many learners still need several run-throughs with the numbers 1, 4, and 2 before they realize that they are working with a cycle. Apparently, these learners are paying much more attention to calculating than they are to observing what is calculated. In such cases, there is a need to explain why when the generating rule repeats an output value, it has arrived at a cycle.

Mathematical follow-up questions

The simplified generating rule for the Collatz sequence I already mentioned can lead to the question, What would the most economical version of this rule look like? You could discuss the following suggestion: First, halve the number n. If the result is a whole number, then that is the next member of the sequence. If it's not a whole number, round this number to the least integer greater than or equal to n and add it to n. For example, if you start with 7 and apply the Collatz rule, you get 22, 11, 34, 17, ... If you use this halving shortcut, you get 7, 11, 17, ... How many calculation steps have been saved when this version arrives at a cycle?

You can visualize all the examples found by your students by combining them into a single graph. This directed graph with the number 1 as the root condenses the orbits of all different starting numbers in one picture.[1]

The concept of this imagining task aims to promote a discussion about the difference between a plausible guess and a mathematical proof. Based on your

---

1   There are many ways to visualize the Collatz sequence, such as https://oeis.org/A008908/graph, https://www.dedoimedo.com/physics/online-search-collatz-conjecture.html, and https://www.jasondavies.com/collatz-graph. For a picture of a *Collatz tree*, see Glaeser and Polthier (2020).

students' experiments, the Collatz conjecture indeed suggests itself: Regardless of the number they start with, the described sequence results in the cycle $\ldots \to 1 \to 4 \to 2 \to 1 \to \ldots$ even if it is a long way to get there.

However, no one has yet found a way to prove whether this is true for all the infinitely possible starting numbers. Even though computer calculations have demonstrated that all Collatz sequences with a starting number of less than 10^{20} end in this cycle (as of 2017), we cannot conclude that this is also true for *any* arbitrary starting number.[2] For this purpose, it's also no help to know the proven fact that, should a Collatz sequence result in a cycle without 1, the cycle would need to have over 100 million steps.

It's natural to want to conclude that a conjecture that is true for so many numbers applies to all numbers. Here are two well-known examples that teach us to be wary of such generalizations:

- The expression $991n^2 + 1$ is not a perfect square for the first 10^{27} natural numbers. Nevertheless, there are numbers for which this expression becomes square. The smallest such number is $n = 12,055,735,790,331,359,$ $447,442,538,767$.

- Pólya conjectured that half or more of the natural numbers less than any given natural number $n > 2$ have an *odd* number of prime factors (counting repeated prime factors). So, for example, for $n = 14$, there are eight smaller integers with an odd number of prime factors $(2, 3, 5, 7, 8, 11, 12, 13)$ and six smaller integers with an even number of prime factors. Pólya had verified his conjecture up to $n = 1,500$. In 1960, though, it was proven that Pólya's conjecture is not correct for $n = 906,150,257$ (which is known to be the smallest counterexample since 1980).

To put it succinctly, a proof can never be replaced by testing many numbers. This imagining task thus refers to a piece of mathematics that, from the mathematician's point of view, is not finished. To be sure, the mathematical question is straightforward to understand and seems to possess a plausible answer. However, we math teachers ourselves have no certainty about the answer, and this is an amazing experience for many of our students! Especially so, when they realize that,

---

2   See http://www.ericr.nl/wondrous/pathrecs.html.

even today, mathematical researchers are still trying to find the definitive answer to the Collatz conjecture.

As a generalization of this exercise, you can also apply the generating rule of Collatz to negative starting numbers. Doing this has little impact on the mathematical situation: the number sequence appears to become cyclic for all negative starting numbers as well. However, instead of having just one cycle, to date three cycles have been found, with lengths $2, 5$, and 18.

You can also modify the generating rule. What happens if you map odd numbers n to $3n - 1$ instead of $3n + 1$? What about if you map odd numbers n to $5n + 1$? A variation that comes quite close to the Collatz rule is: Map n to $\frac{n}{2}$ if n is even, map n to $\frac{n}{3}$ if n is divisible by 3, but not by 2, and, in all other cases, map n to $5n + 1$. The resulting sequence, like the Collatz sequence, appears to arrive at 1 for all starting values and to become cyclic ($\ldots \to 1 \to 6 \to 3 \to 1 \to \ldots$). However, a proof has yet to be found.

Students can ask further questions and generate many hypotheses with the help of computers. What is the maximum value of a Collatz sequence? (For example, the sequence for the starting number 27 takes the maximum value of $9{,}232$.) Which maximum value is taken most often? (The maximum value of $9{,}232$ is taken by over a fourth of all Collatz sequences with a starting number $<2{,}000$.) Why do some pairs of consecutive starting numbers have the same number of elements in their Collatz sequences before arriving at 1 (e.g., 18 and 19, and 340 and 341)?

PS2 SLIDING A LADDER

DIFFICULTY LEVEL	★ ★ ★
MATHEMATICAL IDEA:	The path of the midpoint of a ladder sliding down a wall is a fourth of a circle.
PREREQUISITES:	• Handling a ladder (setting it up, causing it to slide down a wall).
	• Midpoint of a line segment.

The circle as a special plane curve is one of the standard subjects we cover with our students at the secondary level. Generally, we teach the circle as being the set of all points $(x,\ y)$ at a fixed distance r from another point, such as the origin. At upper secondary level, this static conception leads to the Cartesian equation $x^2 + y^2 = r^2$. However, this equation doesn't reflect the fact that the circle can also be described as the path of a moving point. The ellipse is mentioned in connection with plane conic sections, but most high school curricula don't include its algebraic description except in advanced classes where it is taught in the context of second-order curves. In all these cases, we present it to our students as a static object.

In this exercise in Mathematical Imagining, students do not construct the circle with a drawing compass. Instead, they produce it mentally with a sliding ladder. The ladder's midpoint follows a circular trajectory, while the trajectories followed by its other points are elliptical. As challenging as this construction is, it reveals new and broader contexts, and students' explorations help them understand the circle as one particular ellipse in the family of ellipses. Conversely, they can conceive the ellipse as a stretched and hence generalized circle. Moreover, the genesis of circles and ellipses as trajectories highlights their mobile, dynamic aspects. This perspective complements the static view underlying the definition via properties as "the set of all points for which . . . applies."

- Imagine a *ladder* in a light and empty room. . . .

- Take the ladder and lean it *closely* against a wall. . . .

- Imagine yourself at the left-hand *side* of the ladder. Lean your left shoulder against the wall. Now you see only *one side rail* of the ladder leaning against wall in the room. . . .

- A light bulb is attached to the *middle* of the side rail you are facing. *Darken* the room and turn on the light bulb. You see it *shining* as a point of light. . . .

- The *bottom* end of the ladder begins to slide slowly along the floor to the right, *away* from the wall. The *top end* of the ladder continues to touch the wall and slides *down* it. When the top touches the floor, the ladder stops sliding and comes to *rest*. . . .

- What is the shape of *the trace of light* that the bulb "draws" in the darkened room as a result of the ladder's sliding?

- What did you imagine during this exercise in Mathematical Imagining?

Notes on intended mental images

This exercise in Mathematical Imagining begins by describing a mental image: an empty room, a ladder with a light bulb, and an individual who is imagining the scene. It asks students to place the ladder against the wall and position themselves to the side of it as an observer. Then students are asked to imagine the ladder beginning to slide, while keeping an eye on how this movement plays out. The steady sliding of the ladder means that students must make mental movements while constructing the scenario itself and not only when they begin working on the mathematical problem.

While exploring the mathematical question, students should concentrate on the ladder's midpoint and also keep a mental eye on the entire light path it moves along. This challenge requires your students to make repeated shifts in perspective between the sliding ladder, including its surroundings, and a specific object within this scene: between (1) the sliding ladder in its role as the determining factor of the path of light and (2) the light bulb that is moving as a result of the sliding action.

Notes on the productive and counterproductive mental images that learners construct

As we discussed in Chapter 2, this imagining task almost inevitably gives rise to a particular, counterproductive mental image.

- Most students "see" the path of light as a curve tangent to the wall at the beginning and to the floor at the end, and whose form is convex downward.[3] For one of my student's sketches of this, see Figure 2.2 (*top left*).

---

3   A twice differentiable function f of a single variable x is *convex* (or *convex downward*) on an interval if and only if its slope increases on the interval, which means if $f''(x) > 0$. Visually, it "curves up" to the right, like a cup or a hanging rope. If its slope decreases on the interval (which means $f''(x) > 0$), the function f is *concave* (or *convex upward*); it "curves down" to the right, like an arch or a hill (Weisstein 2003, 292, 326).

- The cause of the dominance of this convex light path is most likely that, even though students have darkened the room, they find it difficult to sufficiently suppress the image of the sliding ladder altogether. As a result, the family of ladder positions—"ladder lines"—pushes the family of light bulb positions into the background. Students then mistake the envelope curve of the ladder lines for the path of light. A Gestalt principle is probably at work here. (For a student's analysis of this phenomenon, see Figure 2.5.)

- As the ladder is leaning against the wall, some of your students may be tempted to climb up it. If you are standing on the ladder as it begins sliding, you will likely imagine a much stronger sense of the vertical falling motion (due to gravity) than of the horizontal motion away from the wall. This nonvisual mental image likewise supports the conjecture that the light bulb's trajectory begins tangential to the wall and, as a consequence, might be convex.

- It occasionally happens that a student cannot manage to darken the room, and then they have difficulty recognizing the light path. Further factors that might cause your students' visualization difficulties are how they attach the light bulb to the ladder or how they coordinate their own body and the ladder.

Even if the mental image of the convex curve remains dominant, it is possible for students to modify or counteract it. This revision allows them to develop the following productive mental images:

- As in a thought experiment, students can make the ladder slide and push it back again as often and as fast as they want. As basic as this may seem, it is actually quite helpful, particularly for learners who have little previous experience with exercises in Mathematical Imaginings.

- It is easier for students to observe the beginning and the end of the sliding movement. At the start, the light bulb moves contrary to the sense of falling and runs almost horizontally and, hence, perpendicular to the wall. At the end of the slide, the light bulb movement is almost vertical and, thus, perpendicular to the floor. Therefore, the path of light cannot be at a tangent to the wall and the floor.

- If students are not able to have a clear mental image of the path, they can perform "stop and go" mental actions on the continuous movement: They stop the ladder, take a photo, allow the ladder to slide a little farther, take another photo, and so forth. While this action at first gives students a series of points of light rather than a clear picture of the path as a whole, students can ultimately integrate the individual into a single image.

- Students can approach the mathematical question by testing conjectures, for example, that the light bulb's trajectory is straight. To do this, it helps if they imagine the midpoint of the ladder running along a guide rail with the shape of the conjectured curve (here: the straight line). Can the ladder then continue sliding so that its ends touch the wall and the floor? Or does the straight guide rail block the ladder's movement?

- Instead of a guide rail, students can add a rod joining the middle of the ladder with the place where the wall meets the floor. The mental movement now begins with this rod and not with the ladder. Provided that this rod construction can be moved as the ladder moves, the circular movement is then much more likely to suggest itself. (Recall also Figure 2.8.)

- Students may want to use their hand to model the ladder, with their large knuckles representing the point of light. When they mimic the action of the ladder sliding, they can watch their knuckles to see in which direction they move. This model may help students focus more on the light than on the ladder.

- To counteract the mental image of a convex path of light, students can supplement the described situation. They imagine a second ladder in the room. It is placed next to the first ladder in such a way that its lower end is mounted securely at the meeting place of wall and floor but free to rotate there. Its middle is connected by a pivot with the middle of the first ladder so that the two ladders form a kind of folding ladder in the shape of an *X*. While the first ladder slides down, the second rotates (around its fixed lower end) toward the floor—showing that the light bulb is in fact moving along a circular path. (See Figures 2.6 and 2.7, where a student of mine brought the sliding and the falling ladder together in the image of a pair of scissors.)

Mathematical follow-up questions

Once your students have engaged with the thought experiment this exercise in Mathematical Imagining presents, their next step can be a real experiment. One student slides a yardstick or long ruler along the chalkboard exactly as described in the imagining task. Another student follows the motion with a piece of chalk at the stick's midpoint. The trajectory emerges on the board. The experiment's magnitude and dynamics support the conjecture that the path of light could be convex upward, or concave—and perhaps also circular.

This new conjecture leads to further mathematical activities and questions in two areas: (1) modeling and proving and (2) generalizing and varying.

- In order to prove conclusively that the path is circular, learners create a mathematical model of the situation. They can modify the mental image of the crossed ladders (or pair of scissors) by the quadrilateral determined by the ends of the ladders. Because the two nonright angles in every right triangle add up to a right angle, this quadrilateral is a rectangle. Together with the fact that the two diagonals in a rectangle are of the same length and bisect each other, this implies a constant distance between their intersection and any corner of the rectangle and, hence, the circular form of the path of light.

- Students can also create a model by introducing a circle whose diameter is the ladder, by extending the figure with similar triangles, or by embedding the ladder in a coordinate system. Depending on the models and your students' level of knowledge, you could follow up with synthetic geometric proofs (Thales circle) or algebraic proofs (Pythagorean theorem, equations of proportion).

- You can also slightly vary the question posed by the imagining task: What is the shape of the light's path when the light bulb is attached above or below the middle of the ladder? The benefit of modeling with a coordinate system is evident here, because the corresponding equations of proportion lead directly to the equation of an ellipse. Therefore, you can use this exercise in Mathematical Imagining, like I did, to introduce the topic of "circles and ellipses as trajectories." This introduction enhances lessons in coordinate geometry, where traditionally parametric equations are used only to describe straight lines and planes.

- The fact that students mistake the convex envelope of the family of ladder lines for the trajectory of the ladder's midpoint also raises further questions. What kind of curve are we talking about here? With the help of calculus, for instance, by interpreting this question as an extreme-value exercise, we find that, when the length of the ladder is 1, this envelope curve has the equation $y = (1 - x^{\frac{2}{3}})^{\frac{3}{2}}$ and is therefore an *astroid*. This is an algebraic curve of degree 6, not only the envelope of the family of ladder lines but also the envelope of the family of ellipses produced by points on the sliding ladder.

In a class that is already experienced with exercises in Mathematical Imagining, learners can develop the sliding ladder scenario further. For instance, they attach the light bulb next to the ladder, perhaps on the semicircle above the ladder (Figure 6.1). Which path does the light bulb move on when the diameter of the semicircle slides along the legs of a right angle? As with the question of the sliding ladder, you can also ask this question in the form of an imagining task. It, too, has an unexpected answer: the light bulb moves along a straight line through the origin (where wall and floor meet). This follows from the inscribed angle theorem or can be proved by an appropriate parametrization of the trajectory. What is the slope of this straight line?[4]

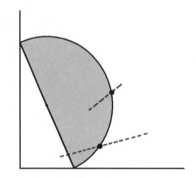

FIGURE 6.1   Sliding semicircle

For all other points lying inside or outside the semicircle and which do not move relative to the sliding diameter, you can ask the questions: Why are their trajectories elliptical? How do you determine the position of the corresponding major and minor axes? You can answer these questions by taking into account the complete circle whose diameter is the ladder.

So far, the two sides along which the ladder slides have always been perpendicular to each other. If you ease this requirement, your students will have another group of questions to explore: Do the points of the sliding ladder create ellipses in this case also? Is there still one point on the ladder that leaves a circular path? And what kinds of trajectories are created by points lying on the circle whose diameter is the ladder?[5]

---

4   The angle of inclination of that straight line is congruent to the angle formed by the following three points: the ladder end on the floor, the ladder end on the wall, and the light bulb (Engel 1998, 319, 324).

5   For a sketch that answers this and further questions, see Glaeser and Polthier (2020).

Producing circles and ellipses using a sliding ladder is much less contrived than you might at first think, as shown by its application in technical devices. The construction of circles and ellipses using a strip of paper is based on the same principle, as is the *trammel of Archimedes*, where two shuttles attached to a rod run in two straight, perpendicular tracks. This instrument makes it possible to draw an ellipse in a single movement (see Figure 6.2).

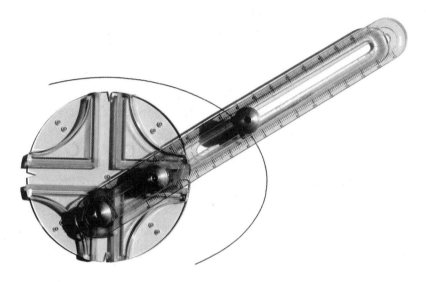

FIGURE 6.2 Trammel of Archimedes

PS3 SLIDING A DRAFTING TRIANGLE

DIFFICULTY LEVEL	★
MATHEMATICAL IDEA:	Suppose a drafting triangle is sitting on two nails so that its right angle points downward. Then this right angle corner moves along a semicircle as the triangle slides along the nails.
PREREQUISITES:	• Drafting triangle as an isosceles right triangle.

hales's theorem has a static aspect that presents the properties of a mathematical subject as well as a dynamic aspect. With it, you can answer the question, "Along which curve do you have to move in order to always view a given object from the same right-angled field of view?" In this light, Thales's theorem is a special case of the inscribed angle theorem.

This exercise in Mathematical Imagining uses another, also dynamic, approach. In it, your students mentally sit an isosceles, right-angled drafting triangle on two nails so that its right-angled corner is pointing downward, and both legs can slide over the nails without the tool falling between them. When one of the upper corners is moved, the other two corners move accordingly. What happens? What, for example, does the trajectory of the lower corner look like?

- Imagine hammering a *nail* into a wall. Then at the same height, hammer a *second* nail into the wall about a handbreadth away from your first nail. . . .

- Now pick up a large 45°-45°-90° *drafting triangle* and hold it so that its right-angled corner is pointing downward and the two 45° angles are pointing to the left and right. Keeping your drafting triangle in this position, sit your triangle on your two nails. Is your drafting triangle sitting *crooked* or *straight* between the nails? . . .

- Now press *down* carefully on *one* of the triangle's *upper corners*. . . . This changes the triangle's position between the two nails. Notice the path that the *lower*, right-angled corner moves on due to the changes. What form does the path have? . . .

- Now pull one of your nails *out* of the wall. Then hammer it back into the wall a little bit *beside* its first position so that you can sit your drafting triangle on the nails again. . . .

- Once again, press *down* on one of the *upper corners* of your drafting triangle. Notice the path of the lower right-angled corner. What form does it have now? . . . How are both curves, the old one and the new one, positioned in comparison *to each other*?

- What did you imagine during this exercise in Mathematical Imagining?

Comments

The lower corner of the drafting triangle moves on a semicircular arc because Thales's theorem applies. This corner, along with the two nails, forms a right triangle whose hypotenuse is the (fixed) segment between the two nails. The diameter of the semicircular arc is then the distance between the two nails.

Mentally, the isosceles drafting triangle and the "Thales triangle" (which is not an isosceles triangle) beneath the two nails easily get in each other's way, especially since both are right triangles. Additionally, when compared to the usual illustrations of Thales's theorem, the described mental image is upside down. This means that even proficient students might have problems recognizing Thales's theorem in their constructed mental images.

With some of your students, their drafting triangle will slip through the nails when its right-angled corner approaches one of the nails. This raises the question of what is the minimal distance required between the nails so that this can't happen.

Mathematical follow-up questions

After you've worked with the mathematical problem of the imagining task, you can vary the instructions and the questions. For example, students can sit their drafting triangle on the nails so that one of the acute-angled vertices hangs downward.

- What happens in this case with the lower vertex, when one of the upper vertices is pressed downward? What does the trajectory look like? Even though the asymmetrical position of the drafting triangle might get in a student's way, here, too, the trajectory is circular.

- What fraction of a full circle does the trajectory consist of? How much longer is it than the trajectory in the original imagining task? As each base angle of the drafting triangle is $45°$, the inscribed angle theorem tells us that the full path consists of a full circle reduced by twice this angle, that is, three-fourths of a full circle. Letting d stand for the distance between the nails, then the diameter of this trajectory circle is $\sqrt{2} \cdot d$. Hence, the trajectory is more than twice as long as that in the original exercise.

- A further question that arises is, "What paths do the two upper corners move on when the right-angled vertex points downward and the drafting triangle is moved between the two nails?"

This question goes far beyond any high school curriculum as it leads to algebraic curves of the fourth order. Nevertheless, we can pose questions about the form of these trajectories the upper corners would follow: Can you guess the form of a curve segment and, based on your mental image, sketch it? If the drafting triangle is moved only a little bit back and forth, the upper vertices appear to follow a circular path. However, this guess does not prove true since here we are actually dealing with *Pascal limaçons*: depending on how the length of the leg l relates to the distance between the two nails d, either the result is a convex curve (if $l \geq 2d$), a curve with a cusp (if $d < l < 2d$), or a curve with a loop (if $l < d$—and for $l \leq \frac{\sqrt{2}}{2} \cdot d$, the drafting triangle should be thought of with extended legs lying on the nails so that it will not fall through them). In the special case $l = d$, you obtain a curve with an inwardly directed cusp (a so-called *cardioid*). For $l \to 0$, the acute-angled corners—as the extended legs slide over both nails—move along a circle doubled over the Thales circle.[6]

---

6 Because the equation of a trajectory in polar coordinates is $r = d \cdot \cos \varphi + l$ with the origin in one of the nails, the equation in Cartesian coordinates is $(x^2 + y^2 - d \cdot x)^2 = l^2 \cdot (x^2 + y^2)$, which is an algebraic equation of degree four.

PS4 LAYING OUT A CHAIN OF SQUARES

DIFFICULTY LEVEL ★ ★

MATHEMATICAL IDEA: Repeated actions can lead to different, yet equivalent, algebraic expressions.

PREREQUISITES:
- Arrange matchsticks in a chain of squares.
- Recognize individual squares within a chain of squares.
- Mental calculation up to 50.

Modern textbooks contain various ways to introduce algebraic concepts. One approach to develop algebraic reasoning is to examine *pictorial growth patterns,* which serve as a context for exploring generalization: because each term in the pattern depends on the previous term and its position in the pattern, the pattern can be generalized and described by an algebraic rule. Last, but not least, students experience that they can calculate with numbers they don't even know just by using "arbitrary x-variable" numbers.

The growth pattern presented here consists of a sequence of squares laid out with matchsticks, arranged in a chain. By describing the growth pattern in their own words, students identify how many matchsticks are needed for any term in the pattern.

- Imagine that you are sitting at a table. There is a box of *matches* lying on the table. . . .

- Open the box: It is full of matches. Now take out *four matches*. . . .

- Lay your four matches on the table so that they form a *square*. Make your square like this: Lay your first match *vertically* on the table in front of you. Lay the next one so that it connects to the *top* of your first match and points *horizontally* to the right. Lay your third match down so it connects to the *bottom* of your first match *horizontally* and also points to the right. Lay your fourth match *vertically* so that it closes the resulting *gap* on the right. . . .

- Now focus on the *square* you have just laid out. . . .

- Now add *three more* matches to your square on the right to make a total of two squares that are *joined together next* to each other. The whole thing looks a bit like two *links* in a *chain*. . . .

- How many matches have you used *in total* for your chain of squares up to now? . . .

- In the same way, lengthen your chain with a *third link*. . . .

- How many *additional* matches did you need for this? And what is the *total* number of matches you have laid in your chain until now?

- Assume that a matchbox contains *forty-six* matches. *How many* links does a chain of squares have, if you use *all* the matches in your box?

- What did you imagine during this exercise in Mathematical Imagining?

Comments

Without deeper reflection, your students can see from their mental images that they needed three matches for each additional link. This means that the multiples of 3 (increased by 1) are hidden in the chain of squares. So students can make a rough estimate of the chain's length (about a third of the matches).

Depending on the way your students mentally structured the growth pattern, they will describe the dependence of the number of links x to the number of matches differently, and the algebraic expressions vary accordingly. When a student describes it as "after the first match, three more matches are added for each link in the chain," the expression will be $1 + x \cdot 3$. When a student says, "The first link of the chain has four matches, and then, for every new link, you add three matches," the expression $4 + (x - 1) \cdot 3$ results. Although the two expressions look different, they are equivalent.

Mathematical follow-up questions

There are many possible variations on this imagining task. You can investigate the dependence of the number of squares on the number of corners, or you can build a chain of triangles instead of squares, and so forth. Beyond such activities, the examination of chains of squares can also lead to *Euler's formula for planar graphs*. The number v of vertices in a net minus the number of edges e plus the number of enclosed areas f is—for a connected plane graph without any edge intersection—always 1, regardless of which decomposition in polygons is chosen for the area enclosed by the graph: $v - e + f = 1$.[7]

---

7   This theorem can be used to prove Euler's polyhedron formula in which a wire-frame model of a polyhedron is projected on the plane (Hilbert and Cohn-Vossen 1952, 290–292).

Students can prove this formula by taking apart their chain one step at a time. When they reduce their chain of squares link by link, the alternating sum doesn't change, because every removed link contributes a net value of $v - e + f = 2 - 3 + 1 = 0$. At the end, one "edge" is left over that has two "vertices" (and no enclosed "faces"), which satisfies the equation $v - e + f = 1$. The corresponding expressions can also be directly compared with each other, without step-by-step deconstruction of the chains: $v - e + f = (2 + x \cdot 2) - (1 + x \cdot 3) + (x) = 1 - x + x = 1$.

PS5   CONSTRUCTING THE SEQUENCE OF SQUARE NUMBER DIFFERENCES

DIFFICULTY LEVEL   ★

MATHEMATICAL IDEA: The differences between consecutive square numbers form the sequence of odd numbers.

PREREQUISITES:
- Square numbers and odd numbers.
- Mental calculation up to ~200.

Your students can find mathematical structures not only in repetitive, geometric patterns (see PS4) but also in number sequences. In this imagining task, your students will discover and explore such a structure in the sequence of square numbers.

The instructions invite students to imagine the first square numbers and then construct the differences between two consecutive squares. The values students get lead them to guess that all odd numbers (greater than 1) will appear. After the imagining task, students can prove this conjecture and also learn more about further characteristics of square numbers.

- Imagine the first few *square numbers* and write them down next to each other: 1, 4, 9, and so on. . . .

- Now take a closer look at the square numbers *1 and 4*. Find the *difference* between these two numbers and write it down beneath them. . . .

- Now look more closely at the *second* and *third* square number. Find the *difference* between them and write it down beneath them. . . .

- Do the same thing for a few *more* square numbers. . . .

- What do you notice about the values you found?

- What did you imagine during this exercise in Mathematical Imagining?

Comments

The sequence of differences has the values 3, 5, 7, and so forth and, therefore, seems to be equal to the sequence of odd numbers that are greater than 1. Does this observation apply in general or only for the numbers the students have investigated?

Your students can answer this question using two small square cards whose side lengths differ by one unit: When you place the smaller square on top of the larger one, an L-shaped piece of the larger card sticks out beyond the smaller one. If the smaller card has dimensions $n \times n$ $(n \geq 1)$ and the larger card has dimensions $(n+1) \times (n+1)$, then the L-shaped figure consists of two long rectangles, each with dimensions $n \times 1$, and a small square with dimensions 1×1. The difference between the areas of the two squares is the area of this L-shaped piece: $n + 1 + n = 2n + 1$, an odd number. Hence, the sequence of differences of consecutive square numbers equals the sequence of odd numbers (greater than 1).

Instead of focusing on the differences between the square cards, you can also take an integrative approach. For it, students should see the card with dimensions $n \times n$ as being n L-shaped figures that have been joined together to form the square. (Here, the smallest L-figure is somewhat exceptional as it is a 1×1-square.) These L-shaped pieces have the areas $1, 3, \ldots, 2n - 1$, which is why it follows that every square number can be decomposed into a sum of odd numbers, $n^2 = 1 + 3 + \ldots + (2n - 1)$. For more visual proofs that the sum of the first n odd integers is a perfect square, see Nelsen (1993, 72–73).

Mathematical follow-up questions

You can expand on the result and have students answer a wide variety of questions. For example, with the help of this this equation, students can make certain mental calculations:

- What is $79^2 - 78^2$? $(2 \cdot 78 + 1 = 157)$

- What is $79^2 - 77^2$? $((79^2 - 78^2) + (78^2 - 77^2) = (2 \cdot 78 + 1) + (2 \cdot 77 + 1) = 2 \cdot (78 + 78) = 312$, a multiple of 8 as is always the case for the difference of two odd square numbers.)

- What is the most compact way to express $78^2 + 78 + 79$? What is the corresponding generalization? $(79^2$, because $n^2 + n + (n + 1) = n^2 + 2n + 1 = (n + 1)^2$.)

When students know the sequence of differences of squares, they can easily draw *parabolas* that are translated with respect to the standard parabola $y = x^2$. To draw it, start at the vertex point $P_0(x_0, y_0)$ of the parabola and construct a sequence of supporting points $P_1, P_2, P_3, \ldots$ as follows: From point P_n, you get the next point

P_{n+1} by translating P_n by the vector $\begin{pmatrix} 1 \\ 2n+1 \end{pmatrix}$ upward and to the right (or by $\begin{pmatrix} -1 \\ 2n+1 \end{pmatrix}$ upward and to the left, respectively. When you interpolate these points, you get a sketch of the corresponding parabola $y - y_0 = (x - x_0)^2$. This process can be generalized to arbitrary parabolas with a leading coefficent $a \neq 1$.

The sequence of square numbers has many further characteristics and structures that you can explore with your students in class. For example, in the sequence, odd and even values alternate regularly, because odd numbers remain odd when they are squared, and even numbers remain even when squared.

Furthermore, did you notice that the numbers $2, 3, 7,$ and 8 are never the last digits of a square number, while $1, 4, 6,$ and 9 occur twice as often as the last digits 0 and 5 do? That is a consequence of the underlying decimal number system. Hence, square numbers with 4 as the last digit arise from numbers with last digits of 2 or 8, since when 2^2 and 8^2 are divided by 10, the remainder is the same: with modulo notation, $8^2 \equiv (10 - 2)^2 \equiv (-2)^2 \equiv 2^2 \equiv 4 \pmod{10}$. Square numbers with last digit 5, however, can arise only from numbers that end in 5 because $5 = 10 - 5$.

PS6 PROJECTING AND RECONSTRUCTING SOLIDS

DIFFICULTY LEVEL: ★ ★

MATHEMATICAL IDEA: Constructing a solid that casts one circular, one square, and one triangular shadow in the three coordinate directions.

PREREQUISITES:
- The shadow of a sphere in orthographic projection is circular, regardless of the direction of projection.

- The orthographic projection of a right circular cylinder can have a shadow that is rectangular or circular.

In order for students to be able to sketch a three-dimensional solid, such as a cylinder, on a piece of paper, it is crucial that they can determine its silhouette, that is, the outline of its projection onto a plane. If we have an oblique projection of a number of stacked cubes and we want to know how many cubes make up the "real" solid form being projected, we're in the opposite situation. Students must be able to reconstruct the solid from its projections. However, such reconstructions are not always possible, and even when they are, they may not be unique. Meeting a question in math class that doesn't have a solution or a clear unambiguous answer could be a new experience for your students. Instead of avoiding this aspect of mathematics, imagining tasks bring such questions into the classroom.

This exercise in Mathematical Imagining uses a sphere as its starting point and first asks students about the solid's outline when projected orthographically onto the xy-plane. Learners then modify their mental image of the sphere so that it casts a square shadow on the lateral xz-plane but still retains its circular outline when projected from above. Finally, the exercise asks students how to modify the solid so that it additionally casts a triangular shadow behind it onto the yz-plane.

- Imagine a *table* in the corner of a room. A *large sphere* is hanging over the table. From *high above*, a *bright light* is shining down on the sphere. . . .

- The sphere casts a *shadow* down onto the table *below* it. What does the shadow look like? What shape is it? . . .

- You are now to modify your sphere as follows. It should keep casting the same shadow onto the table below. However, when lit *from the side*, your solid should cast a *square-shaped* shadow on the wall *to the other side of* the table. . . . How do you have to *modify* the sphere to achieve this?

- Can you modify your solid further, so that it casts a *triangular* shadow *on the wall behind the table* when it is lit from the *front*? . . .

- What did you imagine during this exercise in Mathematical Imagining?

Comments

A right circular cylinder whose height is equal to its diameter fulfills the first two conditions. It casts—depending on its orientation—a circular and a square shadow. If this cylinder is cut with two planes running front-to-back that pass through the top center point of the cylinder and spread out so that they just touch the opposite (left/right) points of the base circle of the cylinder, a wedge form arises whose shadow on the back wall is then triangular and thus fulfills the third condition.

The wedge-shaped cylinder might remind some of your students of a toy where children push blocks of different shapes through holes into a crate. Whether because of their experience of playing with such a toy themselves or for some other reason, some students imagine a solid that is different from the one I described earlier. The question the imagining task poses doesn't have one clear answer, so it can well be that their imagined solid also has the required orthographic projections. We'll come back to this possibility next.

My students frequently mention two imagining strategies that they develop for their constructions.

- In one strategy, they imagine stretched-out "cookie cutters" with the three desired cross sections: circular, square, and triangular. By passing these three cookie cutters in turn and at right angles to each other through a "piece of dough," they "cut" the required figure out of the sphere.

- Alternatively, they take three flat surfaces made from cardboard, with the desired outlines: circular, square, and triangular. They slot these into each other so that they are pairwise at right angles to one another, thus creating a scaffold. They then "fill" this scaffold with a substance (e.g., clay), ensuring that none of it protrudes beyond the scaffolding edges (in the sense that the resulting shape can pass through the cookie cutters above without obstructions).

Both of these strategies can be productive and are useful, even when it comes to constructing various different figures to match the given silhouettes.

Mathematical follow-up questions

Is there actually a solid that has circular contours when projected orthographically onto three perpendicular surfaces but that isn't a sphere? Yes, for instance, a sphere that has a dent that does not affect the great circles responsible for the solid's shadows. This solid is still spherical in certain places while in others, it is no longer convex.

If we change our requirements such that the sought-after solid has *no* spherical surfaces, it's still possible to answer this question with "yes." Some students who use the cookie-cutter strategy may construct their solid as the combination of three congruent, right circular cylinders going through each other pairwise at right angles, a so-called *tricylinder*. The region common to the three cylinders intersecting at right angles forms a new solid (called a *solid of intersection*). It has eight vertices, and its surface is composed of twelve curved patches that are each pieces of the three cylinder surfaces. Unlike the dented sphere, this solid is convex (see Figure 6.3).

Students using the scaffold strategy also may find an answer, this time using three circular disks. When they slot the disks into one another (pairwise at right angles), it produces eight "chambers" of a sphere. Then they fill these chambers with substance and smooth each one over until its surface is convex. Because students can carry out this process in many different ways, they can produce an infinite number of convex solids that have circular outlines under parallel projection in each of the three coordinate directions. Unlike the sphere, these solids can have spines. The solid of intersection of the tricylinder, illustrated previously, has the greatest volume of all solids that can be created in this way (Figure 6.3, *right*).

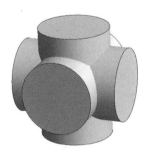 

FIGURE 6.3 Tricylinder (*left*) and its solid of intersection (*right*)

Similarly, there is more than one possible answer to the question posed by the original imagining task. In fact, there are two other known convex solids with the desired properties. To construct them, students begin once more with a right circular cylinder, replacing its top circle with a diameter running front-to-back. From here, the resulting solids differ in the geometry of their lateral surfaces.

- To produce the second solid, bisect the base circle with a left–right diameter that runs below and perpendicular to the upper diameter running from back to front. Then you connect each of the two endpoints of the upper diameter with all of the points of that half of the base circle, which is closer to the respective endpoint. Connect the "inner" points of the upper diameter to both endpoints of the lower one. As a result, you produce a solid that also meets the imagining task's requirements. Its lateral surface—as in the case of the cylinder with wedge-shaped cuts—is likewise partially planar. However, its plane regions are not (semi)elliptical but triangular (formed by connecting the endpoints of the lower diameter with all the points of the upper diameter).

- For the third solid, a straight line slides along the same back-to-front upper diameter used previously, constrained to pass through the base circle while always remaining perpendicular to the upper diameter. This creates the lateral surface of another solid with the desired silhouettes. It is, however, nowhere planar unlike that of the two solids described earlier. The cross sections of this solid perpendicular to the upper edge are triangular, and the cross sections parallel to the plane of the base circle are elliptical. This solid is called a *conoid*.

As the example of these three solids demonstrates, it's not always possible to infer from a finite number of shadow forms the exact shape of the solid body that is casting them. To reconstruct a sphere, we even have to know the shadows projected in an infinite number of directions.[8] The reason lies in the *strict convexity* of the sphere, its roundness: strictly convex solids are never determined by a finite number of shadows.[9] Rather, such bodies can always be altered without affecting any given finite selection of shadow images.

---

8   See Hilbert and Cohn-Vossen (1952, 215).

9   A solid is *strictly convex* when the line segment that connects any two of the solid's points runs completely inside the solid. In other words, the interior of the line segment that connects any two points may not intersect the solid's border—or even touch it.

However, certain convex solids are already determined through specifying a finite number of appropriate shadows. For example, a cube is uniquely determined through the shadows it casts on the three planes parallel to its faces. In more general terms, all prisms can be uniquely reconstructed from a finite number of appropriate shadows.

Artists have long been fascinated by the fact that objects reveal various aspects and appear different depending on the perspective from which they are viewed. The Swiss artist Markus Raetz's work *Hasenspiegel* (*Hare Mirror*, 1988), for example, shows a bent piece of wire that appears as the outline of a hare (see Figure 6.4, *right*) or as a man wearing a hat (*left*), depending on the direction from which it is viewed.

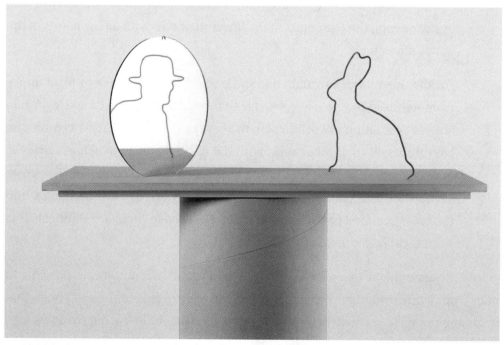

FIGURE 6.4   Markus Raetz, *Hasenspiegel (Hare Mirror)*, 1988

PS7   DRILLING THROUGH A CUBE

DIFFICULTY LEVEL   ★ ★ ★

MATHEMATICAL IDEA: Three tunnels with square cross sections that cross each other pairwise at right angles intersect in a cube.

PREREQUISITES:
- Drill tunnels with square cross sections.
- Retaining your orientation in up to three intersecting tunnels.
- Mental calculations up to 40.

As mentioned in imagining task C2, the cube is likely the most talked-about geometric solid in school mathematics. It is, therefore, suitable for exploring a new class of solids. When students drill through a cube, a solid results that has unusual characteristics—"unusual" in the sense that the solids we normally work with in math class do not have holes.

In this exercise in Mathematical Imagining, you invite your students to imagine a wooden cube. They then go on to connect each pair of opposite faces with a square tunnel that passes through the cube's center. This results in three intersecting tunnels. After students answer questions about the shape and number of the tunnel walls, you can guide them in exploring the shape of the hollow form in which the tunnels cross.

- Imagine that a *cube* made of solid wood is standing in front of you. How many *faces* does your cube have? . . .

- Now imagine that you drill through your wooden cube from the middle of its *front face* straight through its *back face*. Your drill bit drills *square* holes, so the tunnel you drill has a square-shaped cross section. How many *new plane faces* have you created by drilling through the cube's *interior*?

- In a further step, drill into your cube from the middle of its *left face* all the way through the middle of the *right face* using the same drill that produces a *square tunnel*. Your two tunnels *intersect* each other. . . . How many *connected plane faces* does your drilled-through cube have on its *interior* now? . . .

- In the last step, drill through your solid in the same way from the middle of the *top face* straight down and through the bottom face. This third square tunnel passes exactly through the intersection of your first two tunnels. . . . Now look at your cube that you have drilled through three times: How many *plane faces* does your cube have on its *interior* now? . . .

- How many *edges* do your intersecting tunnels have?

- What *shape* does the hollow space in the middle of your cube where you drilled the three tunnels have? . . .

- What did you imagine during this exercise in Mathematical Imagining?

Comments

When students drill through their cubes with the square bit, both square- and cross-shaped faces arise as boundaries of the resulting tunnels. Drilling one tunnel creates within the cube four new faces bounded by a total of twelve new edges. After drilling the second tunnel, there are a total of ten new faces and thirty-six new edges, compared to the original cube. After drilling the third tunnel, there are a total of twenty-four new faces and sixty new edges, compared to the original cube. One way to count these edges is to think there are $6 \times 4 = 24$ edges surround the tunnel openings. Then, each tunnel is divided into three sections: two pieces and the hollow space in between. Each of these three sections has four edges, so each tunnel has $3 \times 4 = 12$ more edges. Tripling to account for all three tunnels, we find $3 \times 12 = 36$ interior edges, and $24 + 36 = 60$. Since each tunnel is drilled parallel to a pair of the cube's faces, the hollow space where the tunnels cross must be cube shaped. This cube arises as the intersection of the three tunnels.

There are three typical imagining strategies that your students may come up with when exploring their drilled-through cube.

- The word *tunnel* often leads to students imagining that their cube is large enough for them to walk upright inside of them, or they shrink themselves small enough to fit. These students answer the mathematical question by walking backward and forward through their first tunnel and looking inside the other two.

- Other students imagine the complement of the drilled-through cube, the drilled-out core. They then can answer the mathematical questions based on this mental image by investigating it from the outside.

- In rare cases, a student will imagine that they rotate the cube. In such cases, the students turn their cube when drilling, so that they drill each new tunnel from front to back. They also turn the cube around in front of themselves in order to answer the mathematical questions.

On this basis, your post-task discussion with students can lead up to the question of the numbers of vertices, edges, and faces of the drilled-through cube. In particular, you can ask them about the number of holes in the drilled-through cube.

Mathematical follow-up questions

Artistically, the German-Dutch artist Ewerdt Hilgemann has worked extensively with drilled-through cubes. For his 1972 installation *Cube Structure*, he produced wooden cubes with differently sized, square-shaped tunnels, as well as their counterparts. Isn't it remarkable that he also depicted the corresponding drill cores? Your students can verify the counts above by exploring Hilgemann's art (Figure 6.5).

Considering Hilgemann's different-sized tunnels can lead to the idea of incrementally varying the width of the three tunnels. Then quadratic functions can be used to answer these questions: Why is the drilled-through cube's surface area largest when the tunnel's width is exactly one-third of a cube side? Which tunnel width makes the drilled-out cube's surface area the same as that of the drill core's?[10]

How many vertices, edges, and faces does the triply drilled-through cube now have? And how many holes does it have? The challenge students face when answering these questions is found less in the drilled holes themselves than in what is left of the original cube faces. Cube faces lose their square shape when they are drilled through, because they each have a square hole (see Figure 6.5, *lower right*). We can decompose the remaining square-minus-a-square faces into smaller, polygonal faces, however, by connecting every corner of an outer square with the corresponding corner of an inner square, leaving us with four, congruent trapezoids on each former cube face. Then the alternating sum "number of vertices minus number of edges plus number of faces," $v - e + f$, can be determined:

- In the case of the cube with one tunnel ($v = 16$, $e = 32$, $f = 16$), $v - e + f$ equals 0.

---

10   The surface area of a cube with three square-shaped tunnels equals the surface area of its drill core if the tunnel's width is $\sqrt{2}/2$ of the cube's width.

FIGURE 6.5   Ewerdt Hilgemann, *Cube Structure no. 140–143* (1972)

- For the cube with two tunnels, it is $v - e + f = -4$, regardless of whether the two cross-shaped faces of the drilled-out core are decomposed into squares ($v = 32$, $e = 72$, $f = 36$) or not ($v = 32$, $e = 64$, $f = 28$).

- With the three-tunnel cube ($v = 40$, $e = 96$, $f = 48$), the result is $v - e + f = -8$.

When compared to a normal cube ($v - e + f = 2$), it's noticeable that the alternating sum is reduced by 2 due to drilling the first tunnel. Drilling through the second tunnel, though, doesn't reduce the sum further by 2 but, rather, by 4. The same thing happens with the third drilled-through tunnel. The reason for this irregularity is because the two (or three) tunnels intersect. If the tunnels didn't intersect but passed by each other, the result would be as expected: $v - e + f = -2$ and $v - e + f = -4$ for cubes with two and three (respectively) nonintersecting tunnels. For such a solid, the *generalized Euler polyhedron formula* holds: $v - e + f = 2 - 2g$, where g is the number of holes in the solid.[11]

Students can also reverse this formula to determine the number of holes. So, for the cube with two tunnels, $g = 3$, and for the cube with three tunnels, $g = 5$. These results become plausible when you imagine that a drill finishes drilling a hole in the moment when the speed of rotation increases again (after it had been decreased due to resistance the wood puts up to the drill). By virtue of this image, students can "hear and feel" that they are making two holes per tunnel when they are mentally drilling the second and third tunnels.

You could also vary this task and have your students use conventional drills with circular-shaped bits that would then result in cylinder-shaped tunnels. What kind of figure results when three cylinders intersect and pass through each other in pairs vertically? How many (curved) faces does this intersection solid have? Is it convex? Does it have ridges? (See the tricylinder in Figure 6.3 for an illustration.)

By the way, as impossible as it might sound, it is technically possible to drill holes whose boundaries are mostly straight segments that are practically square. A drill bit suited for this is shaped like a Reuleaux triangle (see task C7). This bit is milled to provide the necessary cutting edge. To cut an almost-square hole, the bit is attached to a flexible drill chuck that moves within a square jig, synchronized with the rotation of the Reuleaux bit. When drilling, the bit continually touches the jig's four square edges and produces an almost perfectly square hole. The drilled hole does have slightly rounded corners, but its surface area is 98.77 percent of the surface area of a genuine square.[12]

---

11   For a proof of this formula (also known as the *Euler-Poincaré formula*), see Courant and Robbins (1996, 258–259). For the generalization of Euler's polyhedron formula to higher-dimensional polyhedra, see C13.

12   For curves produced by points of a rotating Reuleaux triangle, see Weisstein (2003, 2543–2545), and to see an animated film that shows the bit in action, see http://www.etudes.ru/en/etudes/drilling-square-hole/.

PS8 CURVING A STRAIGHT LINE

DIFFICULTY LEVEL ★ ★

MATHEMATICAL IDEA: The straight line as the locus of all points lying equidistant to two fixed points becomes a hyperbola when one of the fixed points is increased into a small circle.

PREREQUISITES:
- Points that are equidistant to two fixed points form a straight line.

- Increasing a point's diameter above zero produces a growing circle.

Once students have learned about the circle in geometry class, they have had their first—and often their last—experience with a curved shape. Thus, it is hardly surprising that, for learners, circles become the defining example of non-straight curves. Even though the curriculum increasingly takes up curved forms (from nonlinear functions to integral calculus), some students hold onto the notion "if it's curved, then it's a circle." You can use this exercise in Mathematical Imagining to target students' "skewed" perception of curves.

The task begins with students constructing the locus of all points equidistant from two given points: a line. In the next step, they expand one of the fixed points into a small circle. This causes the equidistant locus to curve away from the circle and toward the other fixed point. How does the curve, whose points are equidistant to the circle and to the second fixed point, change as the circle keeps growing?

- Imagine there is a blank sheet of paper on your desk. Now take a pencil and sketch two different points next to each other on your paper, one to the left, and one to the right. . . . Sketch the point that is in the middle *between* your two points. . . .

- Now find and sketch *further* points on your paper that are the *same* distance from your point to the left and point to the right. Do this point by point. . . . What *geometric formation* appears bit by bit? . . .

- Imagine now that your point to the left starts *to expand* and turns into a *tiny little circle*. . . .

- What happens now to the formation consisting of points that have the same distance from your right point and to the little *circle* on the left? . . . What happens if your tiny circle keeps getting *bigger and bigger*? . . .

- What did you imagine during this exercise in Mathematical Imagining?

Comments

The line leading through the two initial points is an axis of symmetry of the entire configuration. Because of this reflection symmetry, students can reduce the complexity of the imaginings by controlling only the upper or lower half of their configuration.

When we have the two initial points, the equidistant locus is a straight line, which is the perpendicular bisector of the segment joining the two points. When the point to the left expands to a tiny circle, the equidistant line must shift a little toward the point on the right. At the same time, it also begins to bend toward this point. This is because the distance of a point P to a circle is defined to be the shortest distance from P to any point on the circle. Students can construct this point by intersecting the circle with the line joining P with the center of the circle. In the imagining task, all such intersection points are to the right of the circle's center, and, therefore, the locus of the equidistant points must also bend, looking like the curve of the letter C.

Considerations such as these can activate the students' misconception mentioned before, that any geometric curve that is not straight must be the arc of a circle. This may explain why students imagine a curved locus as being a circle or at least a circular arc.

Mathematical follow-up questions

Could the locus created in this way actually be a full circle? If this were the case, then the locus would have to be closed, meaning there would be a point where the upper and lower parts of the curve meet. For symmetrical reasons, this point would have to be located to the right of the right point. Here, though, it would be farther away from the circle (around the left fixed point) than from this point, which contradicts its definition. In other words, the locus students are dealing with here cannot be closed and, in particular, is not a full circle.

Still, many students will imagine the locus as a circular arc. This, though, is also not possible. Indeed, the farther a point on the locus is from the symmetry axis, the relatively smaller the circle will appear, and, hence, its pull to the right will also be less. Consequently, the locus is less curved the farther away it is from the symmetry axis, and the locus is most strongly curved at the point on the symmetry axis. Hence, the locus cannot be a circular arc because arcs have constant curvature.

What kind of curve is it then? Based on its construction, the distances between a point on the curve and the two starting points always differ by the same amount, the circle's radius. Therefore, the locus in the imagining exercise is a branch of a *hyperbola* that has its foci in the starting points. If the two foci have the coordinates $\left(-\sqrt{2}, \sqrt{2}\right)$ and $\left(\sqrt{2}, \sqrt{2}\right)$ and the circle's radius at the left point equals $\sqrt{8}$, then the locus is the positive branch of the hyperbola $y = \frac{1}{x}$.

Now what happens when your circle—it plays the role of a directrix—gets larger and thereby approaches the right starting point, or focus? Using the reasoning presented earlier but for a larger circle, we can see that, in this case, the hyperbola's branches do not move as strongly away from the symmetry axis; rather, they are increasingly pressed against this axis. The hyperbola appears to get closer and closer to a *half-line*[13] that starts at the right starting point. In fact, the hyperbola collapses into this half-line in the moment that the directrix circle passes through the focus. Every point of the half-line is equidistant from its starting point, which is, at the same time, the right starting point and also lies on the circle. But other points also meet the equidistant criteria. They are on the line segment between the center of the circle (the left starting point) and the right starting point. The hyperbola as a whole, then, collapses to the half-line that begins at the left starting point and runs through the right starting point on out to infinity.

If students keep growing the circle farther beyond the right starting point, closed loci result. These are confocal *ellipses*, since for every curve point it is no longer the difference but the sum of the distances to the two starting points that is always the same as the circle's radius and, hence, a constant. The two foci of the ellipse are once again the two starting points, whereby its major axis is on the symmetry axis mentioned earlier. The larger the circle's radius becomes in comparison to the distance to the two starting points, the closer the right starting point moves, relatively speaking, to the circle's center. This is why the resulting ellipses tend to become more and more circular. What a metamorphosis to experience: from a straight line to a hyperbola to a half-line to an ellipse to a circle!

---

13   A *half-line* is the subset of a line consisting of a point on the line and all points to one side of the point.

CHAPTER

7

Reasoning Exercises in Mathematical Imagining

This chapter presents you with a number of reasoning exercises in Mathematical Imagining. Each one is designed to make a mathematical idea plausible and credible to learners.

Table 7.1 gives you an overview of all eleven examples in order of their appearance and according to their level of difficulty (more stars denotes more difficulty). I discuss all the imagining tasks in various levels of detail in the comments. In order to demonstrate the educational potential of this kind of exercise, the first two exercises, R1 and R2, include detailed accounts of students' mental images that I have encountered in the classroom. (For the different types of exercises and the structure of the comments, see Chapter 4.)

TABLE 7.1 List of reasoning exercises in Mathematical Imagining

Exercise in Mathematical Imagining	Mathematical idea	Prerequisites	Page
Walking around a triangle ★ ★	The interior angles of a triangle add up to a half-turn.	• Walking around a triangle. • Angle as rotation.	R1 p. 164
Calculating a geometric series ★ ★ ★	The limit of the geometric series $\frac{1}{3} + \frac{1}{9} + \frac{1}{27} + \cdots$ is equal to $\frac{1}{2}$.	• Dividing and coloring a line segment. • Continually repeating an action.	R2 p. 169
Constructing the circumcircle of a hexagon ★	The radius of a circle can be marked off exactly six times along the circle's circumference.	• Laying out geometric figures with equilateral triangles. • Constructing and combining circular arcs.	R3 p. 174
Kites from squares ★	You can make kites by stretching squares in appropriate ways.	• A square standing on one of its corners and its diagonals. • The diagonals in a square are perpendicular to each other.	R4 p. 176
Laying out a special binomial product ★ ★	For positive numbers a and b, $a^2 + b^2$ is less than $\left(a + b\right)^2$.	• Laying out a figure in the shape of the letter L with two congruent rectangles. • Identifying gaps and completing the pattern to form a square.	R5 p. 178
Distorting a triangle along with its attached squares ★ ★ ★	Whether the sum of the areas of the two squares constructed on the legs of an isosceles triangle is greater or less than the area of the square constructed on the base depends on the included angle. In the case of a right triangle, as the Pythagorean theorem tells us, they are equal.	• Equilateral and isosceles triangle with squares constructed on the sides. • Changing the squares by shifting one of the triangle's vertices.	R6 p. 182

(continued)

TABLE 7.1 List of reasoning exercises in Mathematical Imagining *(continued)*

Exercise in Mathematical Imagining	Mathematical idea	Prerequisites	Page
Intersecting spheres ★ ★	The intersection circle of two intersecting spheres arises from a rotation of the spheres.	• The intersection of two circles produces two points of intersection. • A circle rotating around an axis of reflection creates a sphere.	R7 p. 187
Walking around a spherical triangle and turning at its corners ★ ★	On a sphere, there are triangles whose interior angles add up to 270°.	• Walking around a figure on the surface of a sphere. • The sum of a Euclidean triangle's interior angles is 180° (see R1).	R8 p. 191
Traveling around a spherical triangle without turning at its corners ★ ★	When a vector is parallel transported around a closed loop on a sphere, it comes back rotated due to the curvature of the sphere.	• Walking around a figure on the surface of a sphere. • The sum of the interior angles of flat triangles is 180°, and for spherical triangles, it is greater than 180° (see R1 and R8).	R9 p. 194
Circumscribing a square ★ ★	It is possible to circumscribe a square around every closed planar curve that is reasonably "well behaved."	• Rotating a rectangle circumscribed around a fixed shape.	R10 p. 197
Finding the shortest route on the surface of a sphere ★ ★ ★	The straighter a curve connecting two points on a sphere is, the shorter it is.	• Globe with compass directions and equator. • Locations of Boston (USA), and Rome (Italy), on the globe. • Circles parallel to the equatorial circle are circles of latitude. • Planar cross sections of a sphere are circles (see R7).	R11 p. 200

R1   WALKING AROUND A TRIANGLE

DIFFICULTY LEVEL: ★ ★

MATHEMATICAL IDEA: The interior angles of a triangle add up to a half-turn.

PREREQUISITES:
- Walking around a triangle.
- Angle as rotation.

One way to explain the "triangle angle sum" theorem is to tear off the three corners of a paper triangle and arrange them next to each other like three pieces of pie with a common tip. Or, you might rotate two copies of the triangle, each around the midpoint of a different side. Both alternatives illustrate the classical proof first given by Euclid based on the parallel postulate and the alternate angles theorem. These demonstrations shed little light, though, on whether the sum of the angles in a triangle is *exactly* the same size as a straight angle. Isn't it possible to prove this with a single triangle, without mutilating or copying it? Moreover, the standard arguments don't easily generalize to planar polygons, where each additional vertex increases the sum by $180°$.

This exercise in Mathematical Imagining doesn't involve either interpreting the interior angles as static quantities or placing rotated copies of the original triangle next to one another. Instead, we consider the interior angles in their dynamic aspect as angles of rotation. In this imagining task, your students are guided to walk along the sides of a large triangle and, at each vertex, to make a turn equal to the respective interior angle. In this way, the rotations are transported along the sides of the triangle and added together angle by angle.

- Imagine you are walking on a grassy field. You see three long strips of cloth lying in front of you in the shape of a *large triangle*. Stand in the middle of one of the strips of cloth with your nose and toes pointing into the triangle *in front of* you and stretch your arms out sideways to the left and right: they form a line *above* one strip—that is, parallel to and above one side of the triangle. . . .

- Now begin moving *sideways* along the strip of cloth, keeping your nose and toes pointing *into* the triangle as you *sidestep* in the direction of your right arm, placing one foot *next to* the other, until you reach a corner where two strips of cloth meet. . . .

- Your *right* arm now extends beyond the figure, and your *left* arm is above the side of the triangle that you have just sidestepped along. Rotate slowly about your body's axis with your arms firmly outstretched so that your left arm begins to point *into* the figure. Keep rotating your body until your left arm has swept out the corner and arrived *above* the next strip of cloth, and then stop rotating.

- Begin sidestepping along the triangle's side, *this time* in the direction of your *left* arm, placing one foot *next to* the other, until you reach the second corner. Your *left* arm now extends beyond the triangle, and your *right* arm is above the side of the triangle that you have just walked along. Rotate again slowly so that your right arm initially points into the triangle, sweeps out the corner, and arrives parallel to and *above* the next side. . . .

- Continue sidestepping and rotating like this until you arrive back where you started from. . . .

- How are you standing now? . . . What has happened?

- What did you imagine during this exercise in Mathematical Imagining?

Notes on intended mental images

Unlike similar lines of reasoning where you walk forward and backward along the triangle, this exercise in Mathematical Imagining asks students to walk sideways. The advantage of this somewhat unnatural movement is that you actually rotate through the respective interior angle (and not the exterior angle) at each of the vertices.

With their focus on walking and rotating, the imagining task instructions are almost exclusively mental actions in the form of imagined movements. Rather than asking students to manipulate a mathematical object, they encourage students to experience a mathematical fact "physically" with their (imagined!) bodies, through movement. The scale of the intended mental images is therefore larger than in most of the other exercises in Mathematical Imagining.

The exercise gives guidelines for the construction of a series of visual images that direct the mental actions. The outstretched arms serve to point the way when walking and rotating, while the line of sight—given by the direction of the nose and toes—provides an intuitive sense of the accumulation of the rotation angles. Because the exercise encompasses the whole body and prompts movement, students may form mental images that involve their "muscle sense." It's also conceivable that students might have a tactile mental image of their bare feet on the grass.

The closing question, "How are you standing now?" encourages students to compare their line of sight at the beginning and at the end of the exercise. At the beginning, they are facing the inside of the triangle; after having sidestepped around it, the triangle is behind students' backs. The question, "What has happened?" prompts them to form a clear mental picture of the entire route they have taken along the triangle. They should then notice that there are no rotations along the sides, and the rotations at the vertices always turn in the same direction (around the respective interior angle). The combination of these two observations lays the foundations for the "interior angles" theorem for triangles (see next discussion).

Notes on productive and counterproductive mental images that learners construct

As this exercise in Mathematical Imagining involves the body of the imaginer, it presents its own counterproductive mental images.

- Students can tend to lose track of their progress in this exercise. They need to keep in mind how many sides they have already walked along and how many are still to go, and they have to realize that they rotate in the same direction at each corner.

- The instruction to rotate about their own body's axis might tempt students to continue rotating around and around. This might be due to memories of the childhood game where kids rotate about their own axis until they lose their balance and fall reeling to the ground.

In order to overcome these and other obstacles, students develop the following productive visualization aids:

- The instructions include a variety of viewing directions with respect to the triangle. However, these are always from the listeners' local perspective. In order to analyze and answer the mathematical question, it can be helpful if students change their perspective toward the end of the exercise. A bird's-eye view gives them an overview of the triangle and their relation to it.

- Learners should also adapt the triangle's scale as required. While the triangle needs to be larger than the student while they are walking around it, they need to scale down the triangle (and their own body) for the view from above. This involves zooming in and out on the mental images.

- It can be helpful to switch between two different roles. While they are walking and rotating, students are mentally inside their bodies, particularly when comparing their starting and end positions. To analyze the situation from a bird's-eye view, they observe their imagined body from outside.

Mathematical follow-up questions

Imagining facilitates mathematical processes such as reasoning and generalizing. In it, you can guide your students in

- Reasoning about why the sum of the interior angles of a flat triangle is $180°$. The lines of vision at the beginning and the end differ by a half-turn. In order to arrive at this rotation by adding together the individual rotations at the vertices of the triangle, it is necessary to verify that the individual rotations do not neutralize each other but that the turning always occurs in the same direction. As this is the case, the sum of the three interior angles must correspond to a half-turn.

- Allowing them to discover that the interior angle is $180°$. This is true not only for the particular triangle they have imagined here but also for all possible triangles. In short, the sum of the interior angles of any triangle is $180°$, not simply for numerical reasons; it *must* be this angle through the alternation of facing into and facing away from the triangle. Students experience completing exactly a half-turn, pivoting round until they "snap into" place, facing out.

- Generalizing the interior angle theorem to planar polygons and spherical triangles and applying the plausibility argument used here directly to convex quadrilaterals, pentagons, and so forth. Because the view direction at the end of each traverse alternates between "facing into" and "facing away from" when a vertex is added, the overall rotation angle increases by $180°$ for each extra vertex. When students experience facing in and facing out, they see there is no in-between alternative, so they conclude the sum must grow by a half-turn with each additional side. Since we know the angle sum for a triangle is $180°$, this leads to the formula $(n-2) \cdot 180°$ for the sum of all interior angles for a planar n-gon $(n \geq 3)$.

- Making it possible to calculate the sum of the interior angles of concave polygons by applying this method to reentrant vertices (vertices with interior angles greater than a straight angle). It can even be applied to polygons with self-intersecting sides (for pentagrams, see task C5). For the situation with spherical triangles, see R8 and R9.

R2   CALCULATING A GEOMETRIC SERIES

DIFFICULTY LEVEL: ★ ★ ★

MATHEMATICAL IDEA: The limit of the geometric series $\frac{1}{3} + \frac{1}{9} + \frac{1}{27} + \cdots$
is equal to $\frac{1}{2}$.

PREREQUISITES: • Dividing and coloring a line segment.

                            • Continually repeating an action.

Some textbooks reason geometrically to argue that certain geometric series, for example, $\sum_{n=1}^{\infty} \frac{1}{2^n}$, converge rather than diverge. They approach the question from the end and start with a line segment with a finite length—the "limiting value"—which they then exhaust by removing increasingly smaller pieces (Figure 7.1).

0                                      $\frac{1}{2}$ $\frac{1}{2} + \frac{1}{4}$ ...   1

FIGURE 7.1   Convergence of the geometric series $\sum_{n=1}^{\infty} \frac{1}{2^n}$

However, when we are not just interested in whether a series converges but also want to find the sum of a convergent geometric series, we usually use algebra. In the classic proof for the limiting sum, the series $\sum_{n=0}^{\infty} a \cdot q^n$ is "pushed to the right" by one summand by multiplying it with the constant quotient q $(-1 < q < 1)$:

$$a + q \cdot a + q^2 \cdot a + q^3 \cdot a + q^4 \cdot a + \qquad\qquad | \cdot q$$
$$\Rightarrow \qquad q \cdot a + q^2 \cdot a + q^3 \cdot a + q^4 \cdot a + q^5 \cdot a + \ldots$$

When the lower series is subtracted from the upper one, the resulting equation can be solved to determine the sum of the geometric series. In this way, all periodic decimals can be changed into fractions, for example, $0.111... = \frac{1}{10} + \frac{1}{100} + \frac{1}{1000} + ... = \frac{1}{9}$, or $0.999... = 1$.

From the mathematical point of view, this procedure might be elegant; however, it is, in the words of the psychologist Max Wertheimer, "a trick in the sense that it gives no direct understanding of what happens structurally" (1959, 293). In particular, it does not point out that it only can be applied under the condition of convergence. For a divergent series such as $1 + 2 + 4 + \ldots$ (all partial sums are positive), applying this very method would yield the (absurd) value -1.

The instructions of this exercise in Mathematical Imagining go back to an argument by Wertheimer, following his goal of "giving insight into the nature of the series" (1959, 293) and also eventually into the nature of the general sum formula[1]

$$\sum_{n=0}^{\infty} a \cdot q^n = \frac{a}{1-q}.$$

- Imagine you are drawing a horizontal line segment that is about as long as your hand is wide. . . .

- Now, divide your line segment into three sections that all have the *same length*. Color the left section blue and the right one yellow. Leave the middle section uncolored. . . .

- Take a good look at the middle, *uncolored* third of your line segment. Divide it again now into three, equally long sections. Once again, color the left section blue and the right one yellow. . . . Do you see how your *new blue* section and your *old blue* section meet? . . . Notice that your *new* blue section is one-third of a third of the entire line segment, and your *new* yellow section is *equally* long. . . .

- Now keep on *repeating* this procedure, dividing the uncolored middle section into *thirds* and coloring both of the outer sections each time. . . .

- How much of your original line segment would *eventually* be blue? . . .

- What did you imagine during this exercise in Mathematical Imagining?

Notes on intended mental images

To calculate the value of $\sum_{n=1}^{\infty} \frac{1}{3^n}$, the imagining task asks students to keep following the same rule to divide a line segment of fixed length and then color the resulting sections with two different colors. Through this, the entire line segment is increasingly covered by two colored subsegments that always have the same length.

In contrast to many other exercises in Mathematical Imagining, this one doesn't begin with a mental image but, rather, starts with a mental action that leads to a mental image. Through repeated mental actions of dividing and coloring, students work on their mental image. The task instructions describe mental images that are primarily visual, and only the image of measuring the length of the line segment ("about as long as your hand is wide") might also be tactile.

---

1   For several visual proofs of the general geometric sum, see Nelsen (1993, 119–121).

The mathematical question, "How much of your original line segment would eventually be blue?" aims to help students look at the entirety of the mental image they constructed through mental actions so that they compare the length of the blue-colored section with the length of the yellow one and then recognize that the sections are equally long.

Notes on productive and counterproductive mental images that learners construct

To follow the instructions and be able to answer the mathematical question, students have constructed the following productive individual mental images:

- They see the middle section of the line segment as a "stockpile" where they can remove parts from its ends as often as they like, but each time they take away less.

- To help students keep an overview of the task, it can be helpful if they switch back and forth between looking at the whole (the line segment) and focusing on the uncolored sections that will in turn be further divided.

- Particularly productive for answering the mathematical question is the mental image of a balancing act. After each step of "divide and color," the new blue and the new yellow sections are the same size. This is why the two-colored sections are always in balance during the dividing and coloring process.

- A further productive mental image is when students imagine the blue and the yellow sections as growing together steadily. Students can take this mental image so far that they stop imagining the step-by-step dividing and coloring and instead have an image of the two colors evenly flowing in slow motion from both sides of the segment.

During this imagining task, some students develop mental images that, rather than being helpful in the quest for the answer, prove to be counterproductive.

- Some students have difficulties imagining an action that is to be carried out an infinite number of times. This skepticism can be so strong that it prevents them from engaging with the mathematical question. You can try to circumvent such counterproductive images by writing your instructions in the second conditional ("If it were possible, then it would . . ."). I did not take this approach here, choosing instead to let this particular sleeping dog lie rather than awaken it in my students. (More on this follows.)

- When students see a mental image of the line segment that is too materialized—in the form of a chain of pearls of (even very tiny) atoms or a stroke of color—sooner or later they will have problems with the dividing. When individual students start zooming in during the continued dividing of the mental action (and this happens quite often in classroom), it can then happen that the line segment they mentally draw keeps getting wider. This contradicts the fact that a mathematical line has length without any width.

- Occasionally, a student will run into trouble with the prescribed use of the colors blue and yellow. For such students, the colors for the left and right sections can't be freely assigned, as the colors are already determined by their position, like "light green" for left and "dark purple" for right. Such students with synesthesia are better off if you don't assign specific colors to the two sections. Other students, though, will be glad for you to suggest which color to use.

Mathematical follow-up questions

The mathematical potential in this imagining task moves between reasoning, generalizing, and proving:

- You can initiate a discussion about the carrying out of actions an infinite number of times. This activity targets the balance between the blue and yellow coloring. Even though the middle, uncolored section will never be completely used up, after each division step, it provides the same-sized portion for the left (blue) and right (yellow) colorings. Thus, the sections that represent the geometric series will eventually cover half of the entire line segment.

  The "devil" here is in the future tense usage of "will": as already mentioned, this imagining task can stimulate debate over the claim that an action can be carried out an infinite number of times often. This concept, regardless of whether it deals with a concrete action or an imagined one, contradicts our everyday experience of finiteness. In the imagining task, students take a typical, mathematical way (out), which consists of not becoming preoccupied with the underlying difficulties but rather skillfully circumnavigating them. As we can see in the visualization of the convergence of the series $\sum_{n=1}^{\infty} \frac{1}{2^n}$ (see Figure 7.1), the logical argument here deals with a finite line segment and, hence, concerns limiting values. The point here is not to carry out mental actions infinitely but, rather, to catch up with (or even overtake!) the convergence in thought so that students gain a sense of where it leads.

This approach also circumvents problems already formulated in ancient Greek philosophy, for example, the paradox of the athletic Achilles racing against the slow tortoise that had been given a head start. What lends this paradox its special force is the quite plausible extrapolation, based on everyday experience, that an action performed infinitely often must also require an infinite amount of time.

- In a further step, you can have your students divide their line segment into four equally long sections instead of three. They color the section to the left blue, do not color the second section at all, and color the last two yellow. Thus, in the end, a third of the entire line segment would be blue, which explains why $\sum_{n=1}^{\infty} \frac{1}{4^n}$ equals $\frac{1}{3}$. However, students must modify the productive mental image of the balancing act from the case of $\sum_{n=1}^{\infty} \frac{1}{3^n}$, because here there are twice as many yellow segments as blue ones. You can apply a similar argument to make the general sum $\sum_{n=0}^{\infty} a \cdot q^n = \frac{a}{1-q}, (-1 < q < 1)$ plausible.

R3 CONSTRUCTING THE CIRCUMCIRCLE OF A HEXAGON

DIFFICULTY LEVEL: ★

MATHEMATICAL IDEA: The radius of a circle can be marked off exactly six times along the circle's circumference.

PREREQUISITES:
- Laying out geometric figures with equilateral triangles.
- Constructing and combining circular arcs.

The fact that a circle's radius can be marked off exactly six times along its circumference is remarkable. Why does this construction work out so neatly, given that circles are well known in many respects for not fitting well into a Pythagorean worldview?

In this exercise, your students imagine actions, such as laying out, moving, and assembling wooden triangles to produce the figure of a regular hexagon. The fact that every circle can be inscribed with the regular figure of this hexagon means that the radius can be measured out exactly six times along the inside of the circle.

- Imagine an *equilateral* triangle made of wood. Put it on your desk in front of you with one vertex pointing away from you. Move a *ruler* toward the triangle from below until the ruler is lying directly beneath the triangle's lower side. . . .

- Slide your wooden triangle along the ruler to the right by a distance equal to the *length* of the triangle's side. To keep track of the wooden triangle's original position, place a second, identical wooden triangle there. . . . You now have two equilateral triangles next to each other and lined up with a ruler. . . .

- Now, slip a third, *identical* triangle (pointed in the opposite direction) neatly between the two triangles sitting on the ruler. It fits snugly because the distance between the upper vertices of your first two wooden triangles is equal to the length of a triangle side. . . .

- Now remove the ruler and *extend* the shape you just created as follows: place another three equilateral triangles against the shape from below. Two of your new triangles rest against your first two original triangles, and the third new triangle closes the gap between these two. . . . Your six triangles now fit together *seamlessly*. Do you see the regular *hexagon* they form? . . .

- Now place the point of a drawing compass in the middle of your hexagon. Draw a circular arc around each triangle side at the hexagon's outer boundary. . . .

- Look at the six circular arcs. What kind of shape do they form? How do this shape and the boundary of the hexagon lie in relation to one another? . . .

- What did you imagine during this exercise in Mathematical Imagining?

Comments

Following this exercise in Mathematical Imagining, some of my learners reported that they had imagined a slowly turning wheel with six spokes. Presumably partly because of the manner and sequence of the instructions, the curve surrounding the hexagon that some students imagine is often wider than it is high. At times, the individual circular arcs described at the end do not join in a completely smooth line but form a kind of rosette. This may have to do with the fact that students assemble the circumference by joining individual arcs. The midpoints of the circular arcs appear to slip toward the sides of the hexagon (a similar phenomenon occurs in imagining task C7). If so, it can be helpful to mentally remove the hexagon to see the arcs forming a perfect circle. Only those who have wrestled with counterproductive mental images such as these can really appreciate that the regular construction works out so neatly!

The second triangle in the imagining exercise arose by translating the first triangle parallel to its base so that the left corner lands on top of the original right one. The fact that a third congruent triangle slips seamlessly between these two triangles is intuitively clear. For Euclidean geometers, however, it still requires justification. For good reason, we cannot apply the line of thought in this imagining task to the surface of a sphere. A spherical triangle can indeed be translated but not in such a way that all of its points leave behind parallel "lines" as a path. In other words, the fact that all of the points of a Euclidean triangle create parallel line segments of equal length when it is translated is a different formulation of the parallel postulate.

R4   KITES FROM SQUARES

DIFFICULTY LEVEL: ★

MATHEMATICAL IDEA: Kites can be made by stretching squares in appropriate ways.

PREREQUISITES:

- A square standing on one of its corners and its diagonals.

- The diagonals in a square are perpendicular to each other.

In the quadrilateral family (also called the *quadrilateral family tree*), there are a number of special shapes. One of them is the kite, which is usually defined as being a quadrilateral that is mirror-symmetrical with respect to one of its two diagonals. From this definition, it follows directly that the kite has two pairs of adjacent congruent sides and that one of the kite's diagonals bisects the other one.

Another feature of kites is that their diagonals are perpendicular to each other but without being square shaped or even rhombic. This fact, too, is based on their mirror symmetry. However, the kite can be understood more intuitively when it is viewed as a square where two opposite corners have been shifted along the diagonal symmetry line joining them. This approach allows for constructing any possible kite and ensures that right angles between the two diagonals remain intact.

- Imagine a *square.* Now rotate your square until one of your square's corners is facing up, another one down, and one each to the right and to the left. . . .

- Now draw in your square's two *diagonals.* One diagonal runs vertically from the top of your square to its bottom and the other one runs horizontally from left to right. . . . The two diagonals *intersect* each other in your square's midpoint. They intersect symmetrically at a *right* angle. . . .

- Imagine now that you can use the diagonals to *stretch* your square but without breaking it. To do this, take your square's *lower* corner and move it straight *down vertically*, along the vertical diagonal. . . . Do you see your stretched square with its two diagonals? . . .

- Now do the same thing with your square's *upper corner*: take it and move it a little bit *vertically down also* but not too far. . . . Your stretched square looks like a kite now, doesn't it? . . .

- At first, you changed your figure's two *lower* sides, and this made them longer than the sides of your original square, didn't it? . . . But even though the two lower sides are now both longer, they still have *exactly equal* lengths. . . .

- The two *upper* sides are now shorter than they were at the beginning. . . .

- Now take a close look at your kite's *diagonals*: How have their lengths changed? . . . And at what angle do they intersect now? . . .

- What did you imagine during this exercise in Mathematical Imagining?

Comments

For some of your students, the first mental image of a square standing on a corner might seem quite strange. For most students, the classic, upright positioning of a square (with one side on the ground) is all too dominant. (See imagining task C2 for the case of a cube standing on a corner.) This could be the reason that students tend to lose control over their figures' positions during this task. For example, their squares can start rotating.

Mathematical follow-up questions

As soon as your students see the kite as being a suitably stretched square, then you can lead an exploration about which of the square's further properties were kept or lost due to the stretching. For example, in the kite, only one of the two diagonals bisects the other one, and only one pair of opposite interior angles is equal.

Generally, most of us assume implicitly that when we have a quadrilateral whose four sides can be grouped into two pairs of equal-length adjacent sides, we're dealing with a convex planar figure. However, if the upper corner in the exercise had continued to move down until it crossed below the horizontal diagonal, the result would have been a concave kite, also called a *dart*. This doesn't change the fact that, in this quadrilateral, too, the diagonals are perpendicular to each other, but their intersection point lies outside the quadrilateral.

R5 LAYING OUT A SPECIAL BINOMIAL PRODUCT

DIFFICULTY LEVEL: ★ ★

MATHEMATICAL IDEA: For positive numbers a and b, $a^2 + b^2$ is less than $(a+b)^2$.

PREREQUISITES:
- Laying out a figure in the shape of the letter L with two congruent rectangles.
- Identifying gaps and completing the pattern to form a square.

Squaring a binomial results in a trinomial. Trinomials such as $a^2 + 2ab + b^2$ or $a^2 - 2ab + b^2$ are called *perfect square trinomials* because each is the result of squaring a special binomial, that is, the sum $(a+b)^2$ or the difference $(a-b)^2$. Another special binomial product is $(a+b) \cdot (a-b) = a^2 - b^2$. All three special products involve squares and play a decisive role in many sections of high school mathematics. For example, they are used to simplify fractional expressions or in solving quadratic equations without the quadratic formula (*completing the square* to form a squared binomial). We can also make use of them when it comes to converting a fraction with a denominator in the form $a + \sqrt{b}$ or $a + bi$ to one whose denominator contains no irrational or complex numbers.

Accordingly, we can expect dire consequences when our students expand $(a+b)^2$ incorrectly as $a^2 + b^2$. Nevertheless, this happens in our math classes over and over again. It seems as if beginners concentrate primarily on the exponent 2 and simply carry over the additivity of the terms without hesitation. It appears that squaring exerts a kind of suggestive power, whereby the case of $(a \cdot b)^2$ provides a pattern that the students willingly follow, and the multiplicative structure remains.

In order to counter this suggestiveness and guarantee that students can carry out such calculations reliably, learners memorize the special product $(a+b)^2 = a^2 + 2ab + b^2$ or the mnemonic FOIL (for First Outers, Inners Last). To support such efforts and also as proof of the theorem, this identity can be visualized by interpreting the algebraic squaring as the figure of a square and the addition operation as line segments or areas joined to it.

That is, $(a+b)^2$, for positive numbers a and b, is not equal to $a^2 + b^2$ but is instead greater. In order to help students become aware of this difference, this exercise in Mathematical Imagining focuses on the missing term $2 \cdot ab$. Accordingly, the instructions first guide your learners to imagine two identical rectangles, which serve as the materialization of this difference. Then, students arrange the rectangles so that their sides are parallel, and they have exactly one point in common. After that, students will use two suitable squares to complement their constructed geometric figure and turn it into a large square.

By using the two rectangles to construct the complete square figure, the imagining task serves to help your students visualize the often-forgotten mixed term $2 \cdot ab$ of the special binomial product $(a + b)^2$ and also deepens their understanding of the formula. The task is not likely a guarantee that students will from now on correctly square sums of terms, but at least they will have gained the knowledge of what something *isn't* or how something *doesn't* work. This negative knowledge can protect them from repeating this mistake.

- Imagine that you have two *elongated* rectangular wooden building blocks. Your rectangular building blocks are on your desk in front of you, and both of them have exactly *the same form*.

- Now you are going to arrange your building blocks on the table like this: Place one block so that it lies *horizontally* in front of you. Take the second block and place it *pointing away from you* on the table, perpendicular to the first block. . . . Now slide this second block so that it is touching the horizontal block *from above*. Slide this block along the first one, until the two building blocks form the *capital letter* L.

- Now slide your vertical block a little bit to the left. This makes a small *gap* beneath your second block and to the left of the horizontal one. Slide the upper block farther to the left until the two building blocks are touching at exactly one point. Now the gap is *square*. . . .

- Take now a *third* building block: this one is shaped like a square and fits exactly in this gap. Fill the gap with the square-shaped block. . . . Do you see what *form* all your building blocks make together?

- There is *another* gap at the top right between your first two building blocks. Now fill this gap with a fourth wooden building block. . . . What form must it have so that it *fits exactly* in the gap?

- Now look carefully at all four of your building blocks: All together, what form do they have?

- What did you imagine during this exercise in Mathematical Imagining?

Comments

If you have your students sketch their imagined building blocks and then compare their sketches with their classmates', there will be many different visualizations. Additionally, because the concrete scale of the rectangular building blocks does not play a role in the configuration the imagining task presents, the special binomial product is also independent of any concrete proportions. Hence, if your students are already familiar with the algebraic interpretation of geometric objects (rectangles, squares) and operations (addition), then this imagining task as described not only illustrates that $a^2 + b^2$ (the third and the fourth building block) is less than $(a+b)^2$ but also goes beyond this and proves $(a+b)^2 = a^2 + 2ab + b^2$.

The exercise has to do with two apparently quite different realms: algebraic terms and equalities on the one hand and geometric objects and fitting them together on the other hand. Some students might have a difficult time accepting that these realms are in fact equivalent. Especially your more reflective students might insist that they have nothing to do with each other and, therefore, should not be connected. It is helpful not only for these learners but also for all your students when you use visualization aids as a teaching tool to help students begin linking geometry and arithmetic.

Mathematical follow-up questions

After this exercise in Mathematical Imagining, students enjoy applying what they've learned about special binomial products to calculation tricks that not only shorten complicated mental arithmetics but also solve them with impressive speed and reliability. Here are a few examples of such shortcuts for making calculations:

- The square of any two-digit number can be easily calculated when it is reduced to the squares of multiples of 10 and the squares of single-digit numbers. For example, the square of 31 is: $31^2 = (30+1)^2 = 30^2 + 2 \cdot 30 \cdot 1 + 1^2 = 900 + 60 + 1 = 961$.

- Thanks to the square of a sum, we can find the squares of numbers that "end in" 5, such as 35 or 7.5. For example, $35^2 = 30 \cdot 40 + 25$, and 7.5^2 can be calculated as $7 \cdot 8 + 0.25 = 56.25$, because $(n+0.5)^2 = n^2 + 2 \cdot n \cdot 0.5 + 0.25 = n \cdot (n+1) + 0.25$.

In addition to these and other ways of simplifying calculations, the special binomial products also play an important role within the context of number theory proofs.

For example, the difference of the squares of two consecutive natural numbers is always odd because of $(n+1)^2 - n^2 = 2n + 1$. You can also explore this fact with your students with an imagining task (see PS5).

R6   DISTORTING A TRIANGLE ALONG WITH ITS ATTACHED SQUARES

DIFFICULTY LEVEL:	★ ★ ★
MATHEMATICAL IDEA:	Whether the sum of the areas of the two squares constructed on the legs of an isosceles triangle is greater or less than the area of the square constructed on the base depends on the included angle. In the case of a right triangle, as the Pythagorean theorem says, they are equal.
PREREQUISITES:	• Equilateral and isosceles triangle with squares constructed on the sides.
	• Changing the squares by shifting one of the triangle's vertices.

For many people, the Pythagorean theorem is simply the mathematical theorem *par excellence*. After all, it is one of the most frequently mentioned results that comes up in surveys connected to mathematics. In our math lessons, the Pythagorean theorem is mainly applied to calculate lengths and distances, and as a result, it is easy to lose sight of the theorem's intrinsic geometric meaning in the context of rearranging or transforming areas.

This exercise in Mathematical Imagining emphasizes just this aspect of the theorem. The task describes an equilateral triangle and focuses on the squares constructed on the triangle's sides and their areas. What effect does moving one of the triangle's corners have on these squares? With the equilateral triangle, the two upper squares together have an area that is obviously greater than that of the lower square. When you move the upper vertex of the triangle steadily downward, the triangle eventually collapses when the vertex reaches the base line of the triangle. This movement causes the upper squares to shrink. At the end, they have only a fraction of their original areas. Together, this is considerably less than the area of the lower square. Consequently, a position of the upper vertex must exist in which the two upper squares taken together have the same area as the lower square. The Pythagorean theorem stands at this transition point between the so qualitatively disparate situations of a "greater" and a "lesser."

- Imagine a *square* with two horizontal and two vertical sides. . . .

- Now put an *equilateral triangle* on top of your square so that the triangle's base coincides with the square's upper side. . . .

- Observe the two inclined sides of your triangle that run upward from its base. Now construct *squares* on each of these two triangle sides. The triangle is now surrounded by three squares, one attached on each side. These two upper squares look like *wings* sticking out from your triangle. . . .

- Concentrate now on your equilateral triangle with the three *attached* squares: the two upper ones that look like wings and the lower one. You see that the two upper squares together are larger than the square beneath your triangle's base. . . . In fact, they are together *twice* as large. . . .

- Now focus on the upper vertex of your triangle and imagine that your triangle with its attached squares is *animated*. If you move the triangle's upper vertex down a little bit toward its base, then your triangle's two slanted sides—its legs—will become shorter, . . . and, as a result, the two wing squares get *smaller*, right? . . .

- To keep things simple, imagine now that you can only move your triangle's top vertex *straight up* or *down*. This way, the two triangle legs will always have the *same* length.

- Now move your triangle's upper vertex toward its base very *slowly* until this vertex is lying just a tiny bit above the base. . . . The two wing squares shrink in the process, and they come closer and closer together until they almost line up with each other.

- Taken together, your two wing squares are now a lot *smaller* than the square beneath your triangle. What fraction of the lower square are the two wing squares, taken together? . . .

- At the beginning, the combined two wing squares were *larger*, and now they are *smaller* than the lower square. So that means that, in between, there must be a position for the upper vertex so that the two wing squares together have *exactly the same area* as the lower square. . . .

- For this to happen, how *high* above the base does the upper vertex of your triangle have to lie? Is there anything special about the triangle at this position? . . .

- What did you imagine during this exercise in Mathematical Imagining?

Comments

In the equilateral triangle, the angle of the top vertex is acute. The area of the upper squares together is twice as large as that of the lower square. When the triangle's angle is obtuse, in the limit the two upper squares are together only half the size of the lower square. Thus, because of the intermediate value theorem (see notes to R10), there must exist a position at which the two upper squares together have the same area as the lower square.

The transition point where the angle changes from acute to obtuse is exactly the sought-for position. In the very moment when the acute angle becomes a right angle, the area of the two wing squares together is the same as that of the lower square. Here, two copies of the triangle fit into each of the wing squares, while four fit into the lower square. This rearrangement of the surface areas proves the Pythagorean theorem and its converse for the special case of isosceles right triangles.

Mathematical follow-up questions

You can expand this task by letting students explore and generalize the Pythagorean theorem, for example, with questions like: "Does it also apply to other shapes?" "Does it also apply in higher dimensions or to other geometries?" To that end, let me make a few closing remarks.

- The Pythagorean theorem applies not only for squares but also to right triangles constructed on the sides of a right triangle. Students can observe this fact directly. If they divide the right triangle along its altitude perpendicular to its hypotenuse, two smaller right triangles result. When they fold out these new triangles and the original triangle along the original triangle's sides, a triangle with three attached similar right triangles results. Because of their construction, the two triangles on the legs are together equal in area to the triangle on the hypotenuse.

- The Pythagorean theorem applies not only to squares and right triangles but also in general to any shapes constructed on the sides of triangle that are similar to each other. Because of their similarity, their areas always exhibit the same constant ratio to the areas of the corresponding squares or right triangles, respectively.

- A characteristic of a right triangle is that its circumcircle coincides with its Thales circle. This is why a right triangle can be duplicated and rotated by $180°$ to complete a rectangle (Figure 7.2, *left*) with vertices that lie on this circle. In this particular inscribed quadrilateral, the Pythagorean theorem can be restated to assert that, in this rectangle, the sum of the products of the two pairs of opposite sides equals the product of the diagonals. This is *Ptolemy's theorem* and applies in fact to any quadrilateral inscribed in a circle (Figure 7.2, *right*). In other words, the (restated) Pythagorean theorem can be extended to apply to any quadrilateral in which opposite angles add up to $180°$.

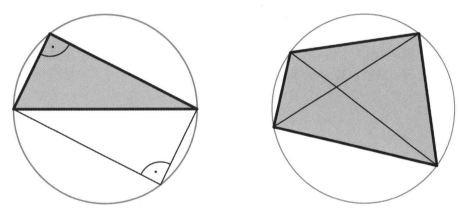

FIGURE 7.2   Ptolemy's theorem (*right*) as generalized Pythagorean theorem (*left*)

- Calculating the length of a diagonal d of a cuboid with length a, width b, and height c results in the "three-dimensional Pythagorean theorem": $\sqrt{a^2 + b^2}^2 + c^2 = a^2 + b^2 + c^2 = d^2$. Thanks to this formula, it's possible to calculate distances in three-dimensional coordinate geometry. The formula can also be transferred to even higher dimensions. Thus, the two corners of a four-dimensional hypercube farthest apart from each other are $\sqrt{1^2 + 1^2 + 1^2 + 1^2} = 2$ length units apart (see comments to C13).

- Pólya (1962, 34–37) described a possibility to generalize the Pythagorean theorem for spatial solids. Cutting off the corner of a cube with a transverse plane produces a nonregular tetrahedron. It is characterized by the fact that all faces meeting at the cutoff corner meet at right angles. When the areas of the corresponding right triangles are labeled with A, B, and C and the cut surface as D (opposite the right-angle corner), the result is not (as might be expected) $A^3 + B^3 + C^3 = D^3$ or $A^{\frac{3}{2}} + B^{\frac{3}{2}} + C^{\frac{3}{2}} = D^{\frac{3}{2}}$ but $A^2 + B^2 + C^2 = D^2$. This theorem is known as *de Gua's theorem*. In a similar way, this three-dimensional generalization of the Pythagorean theorem can be generalized for higher-dimensional solids with a right-angle corner.

- On the surface of a sphere, though, the Pythagorean theorem no longer applies. Although the spherical triangle that covers an eighth of the sphere's surface is a right triangle (in three different ways!), its sides are of equal length a, b, and c. Therefore, it inevitably violates the Pythagorean theorem, $a^2 + b^2 = c^2$. This has to do with the fact that the Pythagorean theorem is equivalent to the parallel postulate or the triangle angle sum theorem: it is exactly the validity of these conditions that decides whether a geometry is Euclidean (see imagining task R8).

R7   INTERSECTING SPHERES

DIFFICULTY LEVEL:	★ ★
MATHEMATICAL IDEA:	The intersection circle of two intersecting spheres arises from a rotation of the spheres.
PREREQUISITES:	• The intersection of two circles produces two points of intersection.
	• A circle rotating around an axis of reflection creates a sphere.

What happens when two different-sized spheres intersect, for example, in the case of two connected soap bubbles? What does their common intersection curve look like? Although in coordinate geometry we may ask our students such questions, the goal is usually not to generate substantive answers but, rather, to make forming and solving the corresponding equations a matter of routine. However, both of these questions can already be answered by middle school students.

In this imagining task, your learners first construct two large, intersecting circles of different sizes and then rotate the circles around the axis of reflection joining their centers. Through the rotation, the two intersecting circles generate two intersecting spheres, and the two points of intersection of the circles generate the circle of intersection of the two spheres.

- Imagine you are drawing a *circle*. . . . Now next to it, draw another, somewhat smaller circle that *intersects* your first circle. . . .

- Now take a close look at the *points of intersection* where the two circles cross each other. Do you see the *two* points of intersection? . . .

- Let's call the figure a double circle. Now with one of your fingers, trace your double circle—but only along the *outside*. . . . Do you feel how the double circle pokes in at the points of intersection? . . .

- Now focus on the *centers* of the two circles. Draw a *line* through the two centers. Both circles lie *symmetrically* to this line. . . . Do you also see that both *points of inter*section lie symmetrically on opposite sides of the line?

- Imagine now how you are rotating your larger circle around the line. . . . For this, you'll need to let your circle leave the drawing plane. Rotate your circle to produce a *perfect sphere*. . . . The same thing happens when you rotate your *smaller circle* around the line. Now you have a second perfect sphere—though, of course, *smaller*. . . .

- Can you now imagine that you are rotating your *double circle* all at once around the symmetry line? What does the resulting solid look like? . . .

- Now please concentrate once more on the two intersection points of the circles. During the rotation of your double circle, they leave a kind of groove behind on the solid. What form does this groove have? . . .

- What did you imagine during this exercise in Mathematical Imagining?

Comments

The groove running between the two spheres is circular. This is because it results from the two points of intersection, which, because of their symmetric position, are equally distant from the rotation axis. The rotation swaps these two points. In a certain sense, this circle represents the seam along which the two spheres are fastened together.

In the same way, your students can construct planar cross sections of a sphere. To do this, they first cut a circle with a secant, which produces two points of intersection. If they then rotate the circle and the secant around the axis that is perpendicular to the secant and runs through the circle's center, a sphere results that has been cut with a plane. In particular, the points of intersection of the sphere and the plane form a circle of intersection, just as in the case of the intersecting spheres above.[2]

Mathematical follow-up questions

Among all solids with equal surface area, spheres have the largest volume. This is equivalent to saying that spheres have the smallest surface area among all surfaces that enclose the same volume. An experimental way to demonstrate this is with soap bubbles floating in air. Because the potential energy, determined by the surface tension, is proportional to the surface area and, in nature, energies are often minimized, the thin soap skin encompasses a fixed volume of air using the least amount of soap film. This explains why the soap film takes on a spherical shape.

---

2   For an animated proof, see É. Ghys, J. Leys, and A. Alvarez, "Chapter 9: The Fourth Dimension," *Dimensions*, http://www.dimensions-math.org/Dim_regarder_E.htm or https://www.youtube.com/watch?v=dI5Zqf20EBc.

Many other properties uniquely define the sphere. For example, the contours and plane sections of spheres are circles—and both properties apply exclusively to spheres. Other properties, though, such as the sphere's constant width and girth, also apply to an infinite number of other solids (see also imagining task C11, and Hilbert and Cohn-Vossen [1952, 215–232]).

When a soap bubble bumps into a planar surface like a pane of glass, it deforms suddenly to a half-bubble. In terms of physics, the boundary condition to encompass a set volume together with a plane results in a half-sphere. Its radius is a good 25 percent larger than the radius of the original spherical soap bubble.

Much more often, though, when two soap bubbles collide with each other and survive the collision, they join and form a double bubble. What does this object look like? Both of the outer, partial surfaces appear to be spherical, too. They are separated by a boundary skin that has a circular boundary. If the two bubbles are the same size, the boundary skin that connects them is planar due to symmetry. What happens when two soap bubbles of different sizes collide? In this case, the boundary skin curves into the larger of the two bubbles since the general rule is that three bubble skins come together at exactly 120°. In order to create this angle, the boundary skin of a double bubble has to curve into the side at which the bubble skin meets the plane at a flatter angle—and this is the larger of the two chambers. In physical terms, there is a pressure difference in the two soap bubbles: the pressure in the larger chamber is lower, which is why the boundary skin bulges into it and not into the smaller chamber.

Your students can take a closer look at how the boundary skin is curved. Could the boundary skin's curvature be constant? If so, then it would have the form of a spherical cap, and the planar surface between two bubbles of equal size would be the limiting case of a spherical cap with an infinite radius (see Figure 7.3).

The conjecture that real double bubbles made of soapsuds consist of three

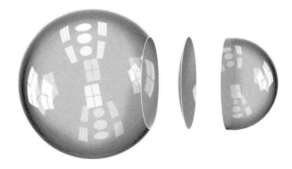

FIGURE 7.3   Double soap bubbles consist of three spherical caps

spherical caps is as old as the investigation of soap bubbles. Surprisingly, though, for "mathematical double bubbles," this was proven only in 2002, in the course of proving that the smallest of all possible soap skins that encloses and separates two different volumes consists of three spherical caps that meet at angles of $120°$ along a circle and, therefore, is a double bubble (see Figure 7.4).

The three surfaces involved in the double bubble are actually spheres. Thus, various quantities associated to double bubbles can be exactly calculated.[3]

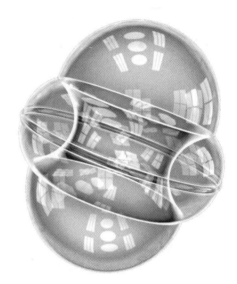

FIGURE 7.4   The double bubble (*left*) needs less surface area to enclose two volumes of differing amounts than any other shape with two chambers (*right*).

---

3   See Isenberg (1992, 89–93).

R8 WALKING AROUND A SPHERICAL TRIANGLE AND TURNING AT ITS CORNERS

DIFFICULTY LEVEL: ★ ★

MATHEMATICAL IDEA: On a sphere, there are triangles whose interior angles add up to 270°.

PREREQUISITES:
- Walking around a figure on the surface of a sphere.
- The sum of a Euclidean triangle's interior angles is 180° (see R1).

The proof that the sum of a planar triangle's interior angles is equal to a half-turn can come alive for your students when they take a suitable walk along the sides of a triangle (see R1). What happens, though, to the sum of the interior angles if, instead, students walk around a *spherical* triangle, that is, a triangle that is drawn on a sphere?

This exercise in Mathematical Imagining describes a mental walk along the sides of a triangle that covers a part of the southern hemisphere. During their walk, students experience that the sum of that triangle's interior angles exceeds that of two right angles. As this task leads into three-dimensional space, it's quite natural for the "walkers" to bring their own bodies into play when making mental movements. In contrast to the walk in the Euclidean plane in task R1, students can carry out the movements in this activity only in their imagination.

- Imagine that you live on a small planet that is as round as a ball. Your planet is small enough so that you can easily walk all the way around it in one day. . . . It also has a North and South Pole, an equator, and the directions north, south, east, and west just like we have here on Earth. Additionally, your planet is made in such a way that you can walk *anywhere* on it.

- Now imagine that you live in a house at the *South* Pole. From there, you are going to go for a walk. Go outside in front of your house, take a look in any direction . . . and now start walking straight ahead to the *north*. Keep walking until you reach your planet's equator.

- Here, take a moment to rest. Then make a *quarter turn* to your right and walk exactly along the equator, step by step, always straight ahead.

- After a while, you are so tired that you decide to take the shortest way back home: You once more make a *quarter turn* to your right and walk straight ahead to the *south* toward your home.

- As you get closer to your house at the South Pole, you notice that . . . you are approaching your house from a *different* direction than the one you took when you left it in the morning.

- Now review your walk. . . . How large is the total angle made by your two turns to the right on the equator? . . . What does the figure formed by the path of your walk look like?

- What did you imagine during this exercise in Mathematical Imagining?

Comments

During the class discussion, students can retrace their experiences with the task on a physical sphere, for example, an apple, an orange, or a ball. Afterward, though, the experience with this imagining task may conflict with students' (already existing) knowledge about the sum of a flat triangle's interior angles. Student reactions like, "This isn't a triangle—it does have three corners, but no straight lines connect them," indicate that learners are more inclined to deny the triangular nature of the figure they have just walked than to broaden their own knowledge about sums of angles. This reluctance to call this shape a triangle may also surface in the discussion about the question to the sum of interior angles in the spherical case. Here, you can hear answers such as "180 degrees—I just don't count the last angle because it is so small."

Mathematical follow-up questions

Once your students have constructed their mental images of a spherical triangle, you can expand the task with further mathematical discussions. When comparing the mental images of the spherical triangles constructed within a class, it becomes noticeable that the sum of the interior angles depends on how far along a student has gone along the equator. The sum of the interior angles is not equally large for all spherical triangles but, rather, depends on the triangle's area, which is yet another striking difference to the plane. If the angles are measured in radians, a spherical triangle on the unit sphere with surface area A has an interior angle sum of $\pi + A$.

Thus, the interior angle sum for a spherical triangle is between π (corresponding to 180°) and 5π (corresponding to 900°).[4]

Asking your students the question of what would happen if they walked around the spherical triangle on the inner side of a sphere also leads to lively debate and many conjectures. A further question for class discussion that could arise is, "Are there surfaces where triangles have an interior angle sum that is less than 180°?"

Some students might come up with the idea of walking along the sides of a spherical triangle in the same way as with the planar triangle in imagining task R1. This would result in a half-turn, just like in the planar case. The next imagining task helps students identify the error in reasoning that leads to this alleged contradiction.

---

4   For comics that introduce spherical triangles, see Petit (1986, 1–19).

DIFFICULTY LEVEL: ★ ★

MATHEMATICAL IDEA: When a vector is parallel transported around a closed loop on a sphere, it comes back rotated due to the curvature of the sphere.

PREREQUISITES:
- Walking around a figure on the surface of a sphere.

- The sum of the interior angles of flat triangles is 180°, and for spherical triangles, it is greater than 180° (see R1 and R8).

Adding together the interior angles of a spherical triangle using the method applied to a flat triangle in imagining task R1 results also in 180°. This seems to contradict the result of exercise R8. So where does the fallacy lie?

If a vector is transported parallel to itself around along a closed loop in the plane, the vector keeps its orientation, and it is pointing in the same direction at the end as it did at the beginning. What happens on curved surfaces like a sphere is quite different. Here, at the end of the motion, the vector is pointing in a different direction than at the start. How much it is rotated depends essentially on the curvature of the surface where the loop "lives." When a vector tangent to the surface of the sphere is transported parallel to itself all the way around a triangle with three right angles, it points in a different direction when it is back at its starting point: The amount by which the tangent vector has been rotated is exactly one right angle. In other words, the vector rotates—despite the local parallel translation—because it has been moved around on a curved surface.[5]

This imagining task invites you on a cross-country skiing tour. If any of your students are not familiar with cross-country skiing, take a moment to explain what it is, perhaps showing a video of cross-country skiing so they have a sense for the

---

5   For an applet that interactively visualizes to what degree a tangent vector does rotate during its parallel transport on a sphere, see John M. Sullivan, "Spherical Geometry Demo," http://torus.math.uiuc.edu/jms/java /dragsphere.

motion. This setting not only makes the difference between the start and end directions obvious but also allows students to verify the vector's local parallelism by noticing the direction of the skis.

- Imagine that you live on a small planet that is as round as a ball. Your planet is small enough so that you can easily walk all the way around it in one day. . . . It also has a North and South Pole, an equator, and the directions north, south, east, and west just like we have here on Earth. Additionally, your planet is made in such a way that you can walk *anywhere* on it. . . .

- Now imagine that you live in a house at the *South* Pole. One morning, you wake up and see that it has snowed heavily in the night, so you decide to go on an outing: You go out of your house and put on your cross-country skis. Under the clear blue sky, you start gliding straight ahead to the *north*. After a while, you reach your planet's equator. . . .

- Take a look over your shoulder at the tracks your skis have left in the snow: There are two *parallel* tracks that run straight behind you to the horizon, . . . where they disappear. . . .

- Look forward again to the north. From this position, begin to step *sideways* to the east at your right. Stomp down with your skis so that they continue to leave tracks in the snow that are parallel to each other. . . .

- After you have been stomping sideways for half an hour, you have had enough. You are still on the equator and you want to take the fastest way back home.

- Keep your body facing the same direction and continue looking to the north: Imagine that the rear ends of your skis also curve upward. . . . Now slide *backward* in the direction of your house at the South Pole. The tracks your skis are making remain *parallel* to each other.

- While you are sliding toward your house, you notice that you are nearing it from a *different* direction than the one you took when you started your outing. Your arrival tracks are *not* parallel to your departure tracks. . . . Somehow or other, you seem to have turned, even though during your outing, *all* of the tracks your skis made were parallel to their neighboring ones! . . .

- Where does this turn come from? . . . Would this turn have happened if you had taken the same outing on a snow-covered flat plateau?

- What did you imagine during this exercise in Mathematical Imagining?

Comments

With this exercise, you can also explain the fallacy mentioned previously. Because a vector in a curved surface already rotates when it is transported parallel to itself, this rotational effect is superimposed on any additional rotations that may have been made. Through the parallel translation of a tangent vector along a spherical triangle with three right angles, the vector is automatically rotated $\pm 90°$. If you rotate it *additionally* at each corner by the interior angle there (three times $\mp 90°$), you get the observed total of $\mp 180°$.

R10 CIRCUMSCRIBING A SQUARE

DIFFICULTY LEVEL: ★ ★

MATHEMATICAL IDEA: It is possible to circumscribe a square around every closed planar curve that is reasonably "well behaved."

PREREQUISITES:
- Rotating a rectangle circumscribed around a fixed shape.

The *intermediate value theorem* states that every continuous function that varies on an interval takes any value between the interval's two boundary values. Although we rarely articulate this theorem in the classroom, the theorem is used as an intuitive fact for a variety of arguments, for example, when calculating the zeros of a function (e.g., the false position method or the proof of the fundamental theorem of algebra) or in proving the fundamental theorem of calculus.

This exercise in Mathematical Imagining also involves using the intermediate value theorem intuitively. It makes plausible that every closed and reasonably "well-behaved" curve in the plane can be circumscribed by a square. In particular, the curve does not have to be convex; it can also be dented, that is, concave.

- Imagine you have a slice of freshly baked bread in front of you: freshly sliced and slightly oval, longer in width than in height. . . .

- Now put a wooden picture frame around your slice of bread so that the bottom of the frame is positioned along the bottom of the bread slice. Your rectangular frame *encloses* your slice of bread so each frame side touches the slice of bread. . . .

- Now look at your framed slice of bread closely. The bread should remain firmly on your desk while you gently start to *turn* the frame around the slice.

- To make this work, your frame has to constantly adjust its size: For a small turn, the lengths of the frame's sides *change*. At first, the two longer sides of your frame will become a little shorter, while the two shorter sides will become increasingly longer. The four sides are still a rectangle that surrounds your slice of bread on all sides.

- Keep turning the frame slowly . . . until you have turned a total of one quarter turn. The frame now encloses your slice of bread just like it did at the beginning. However, now the original longer sides of the frame are the shorter one, and the two original shorter sides are the longer ones.

- It follows, then, that while you were turning your slice of bread, there was a position where all four sides had the *same length*. . . . What did the corresponding alignment of your frame look like?

- What did you imagine during this exercise in Mathematical Imagining?

Comments

Because for rectifiable curves, the rectangle's side lengths vary continuously during the rotation, the intermediate value theorem applies. This theorem, however, just guarantees the existence of a position where the frame has sides of equal length; it does not give a definite procedure for constructing this position. This is why this imagining task asks students to find the claimed intermediate position and form a mental image of it.

It is much more difficult to establish that every simple closed plane curve also admits an *inscribed* square (at least one). Does every simple closed plane curve contain the four vertices of some square? This is also known as the *square peg problem*. As of 2018, it has been proven for curves that are "smooth enough," while the general case still remains unsolved.[6]

Mathematical follow-up questions

According to the intermediate value theorem, simple closed plane curves do admit not only an inscribed square but also an inscribed equilateral triangle. If the curve is also convex, your students can even construct the corresponding triangles.

To carry this out, select an arbitrary point on the curve (the filled-in point depicted in Figure 7.5) and then rotate the original curve at this point twice: first by $+60°$ and then by $-60°$. The two rotated curves each intersect the original curve at a farther point (depicted as a blank point). By construction, the two points of intersection lie at equal distances from the original point, and the corresponding line segments form an angle of $60°$. Thus, the three points are the vertices of an equilateral triangle. As the first point is an arbitrary point on the curve, you get infinitely many equilateral triangles inscribed in the same convex curve.

---

6   For illustrations, see Glaeser and Polthier (2020).

The intermediate value theorem can be used to prove many other results, some quite surprising. For example, you can divide any piece of (two-dimensional) bread that is lying on a plane into four equal pieces by two straight cuts that are perpendicular to each other. A related result is that for any two slices of bread with arbitrary shapes lying on the same plane, there is a straight-line cut that bisects both slices simultaneously. This bisection theorem is also true for any 3 three-dimensional solids, and, in general, for any n bodies in n-dimensions (sometimes called *Stone–Tukey theorem*).[7]

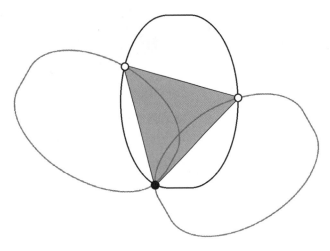

FIGURE 7.5 Equilateral triangle inscribed in a convex curve

The intermediate value theorem can also be of help with a wobbling square table. How can you find a stable position without using shims or shortening a leg? This question can also be posed and answered in the context of an imagining task.

Imagine the table sitting on an uneven floor with its four equally long legs. When looked at more closely, one sees that three of the four legs touch the floor while the fourth is hanging in the air. If this leg is pushed down to touch the floor, then the opposite leg rises up and now hangs in the air. In other words, there is a "wobble axis" running between the feet of the other two legs. If the table is rotated by 90°, then it will wobble just as before because it touches the floor at the same points. The wobble axis, however, now runs between the feet of the other two legs. From the perspective of the table, the rotation has caused it to move from one pair of legs to the other.

If the floor doesn't have any abrupt ledges (like a step), then the intermediate value theorem can be applied. It guarantees that, in the course of the 90° rotation, there must be a position where the table moves from wobbling along the one axis to wobbling along the other axis. At that point, all four feet are on the ground and the table stops wobbling. So to stop a square table from wobbling, rotate it up to 90° around its own axis.

For a further application of the intermediate value theorem, see imagining task R6, which draws upon this theorem in connection with the Pythagorean theorem.

---

7   For details of the proof, see, for instance, Courant and Robbins (1996, 318).

R11 FINDING THE SHORTEST ROUTE ON THE SURFACE OF A SPHERE

DIFFICULTY LEVEL: ★ ★ ★

MATHEMATICAL IDEA: The straighter a curve connecting two points on a sphere is, the shorter it is.

PREREQUISITES:

- Globe with compass directions and equator.

- Locations of Boston (USA), and Rome (Italy), on the globe.

- Circles parallel to the equatorial circle are circles of latitude.

- Planar cross sections of a sphere are circles (see R7).

One of the fundamental experiences in plane geometry is that the shortest way to get from A to B is a straight line. This means that the shortest route never leads over a third point C (unless C lies between A and B). This understanding can give rise to descriptions such as "as the crow flies" to describe the most direct path between two points.

On a sphere, what does the shortest route look like, for example, between two cities located at the same geographic latitude such as Boston, Massachusetts, and Rome, Italy? Relying on their experience in plane geometry, students expect that the shortest route between these two cities coincides with the circle of latitude, probably since these circles usually show up on maps as horizontal lines. The correct answer—that on a sphere the shortest path follows a *great circle*—can cause uncertainty. After all, aren't great circles characterized as being longer than any other circle on the sphere? Or, to put it conversely, doesn't the shortest connecting route between two locations at the same latitude have to actually run along a line of latitude, as circles of latitude are always shorter than the equatorial circle? In other words, your students' mental images and existing knowledge that can be productive in plane geometry can prove counterproductive in the case of the sphere.

However, the following principle of planar measurement *can* be transferred to the sphere. The less curved a line between two points is, the shorter it is. This is why this imagining task focuses not on the length but on the *curvature* of the connecting path between two locations on the sphere. Students will slice through the globe with a planar cut that goes through both cities and the Earth's center. The resulting planar cross section is a circle. Because the cut passes through the Earth's center, it divides the globe in half, which is why the cross section is a great circle like the equator. Consequently, this circle's curvature is less than that of the circle of latitude determined by the two locations. This also explains why the route along this circle is more direct than the one along the circle of latitude. At the end of the task, the students are asked why this is the shortest of all possible connecting routes.

- Imagine that you want to fly from Boston, Massachusetts, to Rome, Italy. For this visualization, imagine that the Earth is a hollow sphere, or globe. On the globe, both of these cities lie equally far north, above the equator. . . .

- To fly from Boston to Rome, your plane can take different routes. The *simplest* route is probably just to fly *exactly to the east* with no veering northward or southward, following a circle of constant latitude. . . .

- Now take a close look at the Earth's *equator*: It is actually a circle, and, therefore, you can imagine it as the boundary of a *circular disk*. Do you see how the center of this circle lies at the *center of the globe*? . . .

- *Parallel* to this disk, slice *straight through* the globe at the height of Boston. . . . The resulting cross section is a circle of latitude, with Boston and Rome located on it. The suggested flight route to the east runs along *a segment* of this circle. . . . Now, put the globe back together. . . .

- Actually, you can choose another, *shorter* flight route: To find it, slice your globe *differently*. Slice with a plane that goes through both of the cities *and the globe's center*. This cut isn't parallel to, but rather, *inclined* to, the Earth's equator. . . .

- Do you see how your cut divides the globe into two halves? . . . Do you also see that the cut circle is exactly *as long as* the Earth's equator and that it also has the exact *same curvature*? And remember: it connects Boston and Rome.

- In particular, it is definitely *less curved* than the circle the first route flew along. . . . *Which* of the two routes would be *shorter* then? . . .

- What would be the shortest route between Boston and Rome, out of *all possible* routes? How do you know?

- What did you imagine during this exercise in Mathematical Imagining?

Comments

As opposed to other exercises in Mathematical Imagining, there is no need for your students to refer to a theorem like the intermediate value theorem (see R10). All they need to know is that great circles are, among all circles on a sphere, those with maximal length but minimal curvature. So, on a sphere, the shortest path between two points—a so-called *geodesic*—follows a great circle.[8]

To answer the mathematical question, it can be helpful for your students to reconstruct tactile mental images and visualize these. To do this, they imagine using their thumb and index finger to make a ring, representing a circle on a sphere. Then, with the other hand, they place a tennis ball on top of the ring. If they carefully press down on the ball so that it starts going through the ring, the ring widens, which means they must keep stretching their thumb and finger farther apart. In the moment when the tennis ball plops through the widened ring, their finger and thumb are less curved than at any time before. Further, this is when the ring and a great circle on the ball coincide.

This less visual, more tactile mental image can be applied directly to the two cities on the globe. First, imagine the globe to be sized so that the tips of your thumb and index finger are touching the two cities while the arc between stretched-out thumb and index finger follows exactly the circle of latitude between the two cities. This allows you literally to touch the mathematical idea of this imagining task. If you rotate your hand slowly around the straight line joining these two fingertips, toward the north, the arc sticks out from the globe; it is too long. This allows your students to experience and to conceptualize that, among all circles that can be drawn on a sphere, the great circle is the one with the least curvature.

---

8   To watch a silent animated film about the shortest route between Moscow and Vladivostok, see http://www. etudes.ru/en/etudes/geodesic/.

C H A P T E R

8

Paradox Exercises in Mathematical Imagining

This chapter provides you with examples of paradox exercises in Mathematical Imagining. While reasoning exercises seek to make a mathematical idea credible to your students, the purpose of paradox exercises is to confront learners with an (apparently) inescapable contradiction. Each exercise is designed around a conflict between different kinds of knowledge (intuitive, experiential, mathematical), which students then investigate and expand.

Table 8.1 overviews the five example tasks. I discuss each exercise at various levels of detail in the comment sections. In the notes to tasks P1 and P2, I share with you in more detail my experiences with mental images that students may form, because mathematical paradoxes are often accompanied by difficulties in forming appropriate mental pictures (in the table, more stars indicate more difficulty). I also suggest possible mathematical follow-up questions for all of the exercises. (For the different types of exercises and the structure of the notes and comments, see Chapter 4.)

TABLE 8.1 List of paradox exercises in Mathematical Imagining

Exercise in Mathematical Imagining	Mathematical idea	Prerequisites	Page
Accommodating guests at the infinite hotel ★ ★	"Calculations with infinity" obey different laws than calculations with whole numbers do.	• Natural numbers on the number line. • If the finitely many rooms of a hotel are occupied, it cannot accommodate any new guests.	P1 p. 205
Excursion into space ★ ★ ★	It could be that the universe is not infinite, but finite.	• Line and plane, circle and sphere (locally and globally). • If you move straight ahead along a line or plane, you never return to the point where you started; this is not true of the same movement on a circle or the surface of a sphere.	P2 p. 210
Slinging a ball attached to a strap ★ ★	Whether and how an object moves depends also on its reference system.	• If you spin around while holding a strap attached to a ball, the ball is forced onto a circular trajectory.	P3 p. 216
Comparing the number of points on and the lengths of circumferences ★ ★	The points of two circles can be matched one-to-one with each other even if the circles are not the same size.	• Turning one clock hand on two circular, concentrically arranged clockfaces of different sizes. • If two line segments consist of a finite and equal number of sections of equal length, they are also of equal length as a whole.	P4 p. 218
Turning the universe inside out ★ ★ ★	The universe might be made in such a way that the length of an object changes when the object moves.	• Interior surface area and interior volume of a large hollow sphere. • The series $\frac{1}{2} + \frac{1}{4} + \frac{1}{8} + \dots$ converges to 1 (see R2).	P5 p. 221

P1 ACCOMMODATING GUESTS AT THE INFINITE HOTEL

DIFFICULTY LEVEL:	★ ★
MATHEMATICAL IDEA:	"Calculations with infinity" obey different laws than calculations with whole numbers do.
PREREQUISITES:	• Natural numbers on the number line.
	• If the finitely many rooms of a hotel are occupied, it cannot accommodate any new guests.

The phenomenon of infinity comes up in geometry in the context of parallels that do not intersect at a finite point. Calculus draws on the observation that there is no largest natural number: $n \to \infty$ represents the process whereby n increases beyond any (finite) bound. In both cases, infinity stands for a concept that we can always get "closer to"—without ever reaching it. This is why we also speak of *potential infinity*.

This exercise in Mathematical Imagining is concerned with a different conception of infinity. Here it is "complete," an entity in its own right. This conception of infinity—*actual infinity*—comes from set theory, where every natural number is paired with a set with just that number of elements. Because sets can likewise be made up of more than finitely many elements, the concept of infinity can be interpreted accordingly. Thus, on the one hand, your students' understanding of numbers acquires an axiomatic foundation, while, on the other hand, it becomes possible to "calculate" with this actual infinity. What happens when adding a (finite) number or when multiplying by a (finite) number?[1]

Our everyday experience—and hence our intuition—tells us that a hotel with a finite number of rooms is unable to accommodate anymore guests when all of the rooms are occupied. This exercise in Mathematical Imagining recounts the paradox of *Hilbert's Hotel* (going back to the German mathematician David Hilbert), in which there is no "last" room. Several groups of new guests arrive. As your students imagine the scene, they are introduced to a paradoxical situation, because the statements, "All of the rooms are occupied" and "No more guests can be accommodated" are no longer equivalent for a hotel with infinitely many rooms. When finitely many

---

1   Here, we are dealing only with countable sets, not with uncountably infinite sets.

new guests arrive, the contradiction is resolved in this imagining task as follows: it is possible to create room for any finite number of new guests by shifting the existing guests to other rooms. However, the question of how a countably infinite number of guests can be accommodated in addition to those already checked into the hotel is left up to your students.

- Imagine a huge hotel: It has *infinitely* many single rooms. These are numbered consecutively with the room numbers 1, 2, 3, and so on. . . . Now imagine that you are working the night shift in this hotel as a *receptionist*. . . .

- One night, *infinitely many guests* are staying in the hotel. You are looking forward to a quiet night and so you close your eyes to take a little cat nap. . . .

- But then a guest turns up at the reception desk, wakes you up, and asks for a single room. "I'm terribly sorry," you say, "I'm afraid we're *completely* booked out."

- "That's not possible," the guest replies. "You have *infinitely* many rooms, so there must be *one* vacant room left for me!" . . .

- "I'm really very sorry," you say, "but we already have *infinitely many* guests staying here, so there is *not even one* vacant room."

- But the guest won't take no for an answer: "I don't believe you. I want to speak to the manager!"

- So, you go and get the hotel manager. She listens to the new guest's complaint and then says to you: "This problem can be solved. We just have to swap the rooms around a little. Move the guest in room 1 to room 2, the guest in room 2 to room 3, that one to room 4, and so on. Each guest moves into the room with the *next highest* number. Room 1 will then be vacant, and so we can accommodate our new guest." . . .

- No sooner said than done. As this is a mathematical hotel, it is all taken care of in the twinkling of an eye. Pleased with a job well done, you continue your catnap. . . .

- A little bit later, though, you are awoken again. A bus carrying a *hundred* people has arrived, and all of them are insisting on having their own *single room*. You have no idea what to do, so you go to get the manager again.

- She says: "We can accommodate these new arrivals, too. This time, move the guest in room 1 to room 101, the guest in room 2 to room 102, and so on. Each guest is moved one hundred rooms *farther*." . . .

- Once again, all of the existing guests have quickly moved and the new arrivals are accommodated in the first hundred vacated rooms. . . .

- But your peace and quiet are disturbed again in the same night. An hour later—it's already long past midnight—an *immeasurably vast* crowd of people starts pouring into the hotel lobby, and their travel guide wants you to give each of the travelers a single room. . . .

- You ask: "How many people are there?" "*Infinitely many*," says the guide. . . .

- What do you, the receptionist, do? . . . How can you accommodate these *infinitely many* new guests in the hotel? . . .

- What did you imagine during the exercise in Mathematical Imagining?

Notes on intended mental images

In order to accommodate the infinitely many new guests, one option is to have the existing guests each move into a room whose number is twice as large. All of the odd-numbered rooms are then free and can accommodate the new guests.

The instructions are geared toward the visual mental image of a hotel with infinitely many rooms and the mental images of finitely and infinitely many new guests. The desired accommodation of one or one hundred new arrivals is solved by the exercise in Mathematical Imagining, which suggests the mental actions of relocating or shifting all of the existing guests. The mathematical question at the end of the imagining task represents a particularly difficult challenge for the mental images formed, as they are inadequate to deal with the task of accommodating a further infinitely many guests in the infinitely many rooms that are already occupied.

Notes on productive and counterproductive mental images that learners construct

To accommodate infinitely many new guests, your students need to modify the mental image of shifting the guests. This often proves difficult. In addition, the following counterproductive mental images may come up in the course of this imagining task:

- If learners visualize the hotel as having various corridors, wings, and multiple floors, the process of relocating guests becomes vastly more complicated.

- Other counterproductive mental images have to do with coordinating the relocation of the existing guests. For instance, when moving into room 2, guest number 1 might bump into guest number 2 (with the same thing then happening with all the other guests). It can be helpful here to ask all existing guests to move at the same time (perhaps with a loudspeaker announcement, "Please move now!").

- How your students visualize the final, infinitely large group of new arrivals can also cause them problems. If, for example, learners attempt to enlarge the hotel lobby accordingly, their attention is diverted from the infinitely large number to the infinitely large lobby, and this distracts them from the issue at hand. It is more helpful to imagine a finite part of the group standing crowded together in the lobby, while the other, infinite part waits outside the entrance.

The following mental images can be useful and productive when carrying out the instructions and answering the questions:

- The exercise in Mathematical Imagining puts the listeners in the role of the night-shift receptionist in order to confront them with the problem. However, it is helpful when students depart from this perspective from time to time to understand the problem and be better able to work on it. Some learners slip into the role of the hotel manager to help find the answer.

- The rooms, much more so than the guests, are crucial to this exercise in Mathematical Imagining. The infinite number of these rooms in Hilbert's Hotel can be captured with the mental image of a long corridor of rooms that appears to grow narrower the farther it recedes into the distance. The listener—in their role as the receptionist—looks down the corridor and can visualize that moving the existing guests takes place through "shifting" them farther along toward the infinitely distant vanishing point. This creates room in the front area for the finitely many new arrivals.

- The room numbers mean that each guest is assigned exactly one natural number. Students might therefore imagine the hotel as a number line running from left to right with points corresponding to rooms. In this mental image of a line of rooms stretched out from left to right, the first two relocation efforts can be understood as mental movements (the simultaneous shifting of all existing guests) to the right. Unlike the image of the corridor of rooms, which reflects the visual perspective of the receptionist, the focus here is less on the newly created space than on the process of moving as a single action.

- To answer the mathematical question, it can help when students modify the mental image of a line of rooms. Parallel to the line of rooms, or next to the existing corridor of rooms, they could add a second series of rooms. The rooms are renumbered in the obvious way, with the existing guests being lodged in rooms whose numbers are now twice as large. The new arrivals can then take shelter for the night in the rooms with odd numbers. Some of my students have talked about imagining a zipper, because the rooms occupied by the existing and new guests slot into each other like the two rows of teeth on a zipper.

Mathematical follow-up questions

Because the infinitely large set of existing guests, represented by their room numbers, can be placed in a one-to-one correspondence with the set of natural numbers, calculating rules such as $x + 1 \neq x$ and $2x \neq x$ no longer apply for the number of guests. Building on the mental images constructed in this exercise, students learn about differences between finite and actually infinite sets and get an idea of the concept of actual infinity. This leads to the following questions:

- How can countably infinitely many buses, each with countably infinitely many passengers, be accommodated in Hilbert's Hotel? One possibility involves sending the existing guests from rooms with the number n to rooms with the number 2^n. The new guests from bus m can then be put up in the rooms whose numbers are the powers of the $(m + 1)$-th prime number. Moreover, in this way (countably) infinitely many rooms remain vacant, such as room 6, whose number is not the power of a prime number.

- How must the guests be assigned to the rooms at the outset so that no one has to move, regardless of whether finitely or infinitely many new guests need to be accommodated later? The new guests should be able to arrive successively, any number of times, every day and every night, and yet no guest should ever have to move because of any new guests.

P2   EXCURSION INTO SPACE

DIFFICULTY LEVEL: ★ ★ ★

MATHEMATICAL IDEA: It could be that the universe is not infinite, but finite.

PREREQUISITES:
- Line and plane, circle and sphere (locally and globally).

- If you move straight ahead along a line or plane, you never return to the point where you started; this is not true of the same movement on a circle or the surface of a sphere.

Looking up at the sky on a clear night, we feel we could fly into the infinite depth of the universe. Based on our local perspective and experiences, we assume that the surrounding, three-dimensional universe is infinite. But can we really be a hundred percent sure that flying straight ahead away from Earth means inevitably to become "lost in space," never, ever to return?

In this exercise in Mathematical Imagining, your students explore, dimension-by-dimension, what consequences a world's being finite or infinite would have on a straight-ahead journey. First, they explore the one-dimensional universe of a giant rope with open ends and then with the ends knotted together. Learners try out which movements are possible in these one-dimensional worlds. Then the task guides them to begin a journey and always travel "straight ahead." When they travel on the closed rope with knotted ends, they will always arrive back at the place they started, even when the rope is very, very long. In the second and third parts of the imagining task, students investigate the analogous questions in higher-dimensional universes that are open and closed in extent, respectively.

This exercise in Mathematical Imagining shows that our three-dimensional universe could feasibly be closed and thus finite. Should that be so, then a straight-ahead flight away from the Earth would not end up lost in the far reaches of space but would someday return. Paths of light wrapping around the finite universe in this way could create the illusion of an infinite universe. To put it more provocatively, a faint speck of light in the night sky could be our own Milky Way galaxy and, within it, our own Earth rotating around our own sun.

- Imagine that your world is that of a *straight line*. You live on a very long rope that is tightly stretched. . . . On your rope, you have only few possibilities of moving: move a little bit *forward* on your rope and then move a little bit *backward*. . . .

- Additionally, imagine that your rope is *infinitely* long. Begin to move in one of the two directions. Keep moving farther and farther. You will certainly never return to your starting point again. . . .

- Now imagine that your rope is not infinitely long, but just very long. Its ends are joined together, and your rope forms a huge *circle*.

- Begin moving on your rope in one of the two directions. . . .

- What happens if you keep moving *long enough* in the direction you took? Will you eventually *return* to the place where you started, or not?

- Imagine now that you are a creature that lives on a two-dimensional surface, like a shadow does. Move in your two-dimensional world, first a few steps forward and then a few steps backward. Then take a few steps to the left and then a few steps to the right. . . .

- Imagine, additionally, that the surface where you live is really *huge* and consists of an infinitely large plane. . . .

- In your flat world, begin now to move on in one direction, keep going farther and farther. You will certainly *never again* return to the point where you started. . . .

- Now imagine that your flat surface is still huge, but it is slightly curved and forms the *surface of a gigantic sphere*. . . .

- Move now in a *fixed* direction on the surface of your spherical world, keep going *farther and farther*. . . .

- What happens if you keep moving *long enough* in the direction you took? Will you eventually *return* to the place where you started, or not?

- We all live in a *universe* where we can move not just to the left and right or, independently from that, forward and backward, but we can also move up and down. . . .

- Usually, we assume that our universe is *infinitely big*: Imagine that you are taking off in a spaceship headed into outer space. You are always flying exactly *straight ahead* away from the Earth, so you will never, ever come back to Earth. . . .

- To conclude, imagine that our universe is *closed*, just like your rope that was curved into a circle and the curved surface of your sphere were. . . .

- Take off again in your spaceship and fly into space, always fly *exactly straight ahead* on a fixed course, keep flying and flying. . . .

- What happens, if you keep flying *long enough* in the direction you took? Will you eventually *return* to the place where you started or not?

- What did you imagine in this exercise in Mathematical Imagining?

Notes on intended mental images

The three parts of the instructions are all structured analogously. The first intended mental image in each part is a locally one-, two-, or three-dimensional world. Within the respective world, your students carry out a small mental movement and explore only a small section of their constructed world. Based on their local experience only, they cannot judge whether this world in its entirety is finite or infinite. Then the mental images of the three worlds are further explored by "moving straight ahead" (meaning following the shortest path). By this mental action, students expand their mental images in order to experience their respective worlds not only locally but also globally. For the locally one-dimensional circle (S^1) and the locally two-dimensional sphere (S^2), respectively, this is not a problem as students can view both worlds globally, from the outside: S^1 as the boundary of a circular disk can be embedded in two-dimensional Euclidean space $\mathbb{R}^2$, and S^2 as the boundary of a ball sits in the three-dimensional space $\mathbb{R}^3$. The third section of the instructions invites students to go one dimension further, which leads to a *hypersphere* S^3. The wish to examine the hypersphere in an analogous way from the outside, however, leads to the attempt to imagine it in four-dimensional Euclidean space $\mathbb{R}^4$, which is a genuine challenge. It can be discussed as a follow-up question (see the corresponding section that follows).

All three sections end with the mathematical question of the possibility of returning. If students have developed the corresponding mental images of S^1 and S^2 (be they locally or globally), they will answer the questions correctly, that yes, they will return. In the case of the hypersphere S^3, they will—at least reasoning analogously to the two lower-dimensional situations—guess that returning is also possible here. In doing so, they contradict the conception that a journey into the universe that leads them away from their point of departure will never lead them back to the starting point again.

Notes on productive and counterproductive mental images that learners construct

As mentioned earlier, this is a difficult imagining task, as it confronts students with some (unexamined) conceptions and imaginative challenges. Notwithstanding, students do construct many mental images. If these images interfere with students constructing and exploring further mental images, they are counterproductive.

- Students' mental images of the one-dimensional world are sometimes accompanied by imagining that their rope is in cold, windy outer space. This is a particularly unpleasant mental image.

- Trying to imagine how the Earth (that looks like a little sphere S^2 when viewed from the outside) is embedded in the universe of S^3 may bring back the image of the sphere S^2 from the second section of the task. This mental image makes it more difficult to take the step into the next dimension.

- The stimulus to imagine a hypersphere in order to decide whether the spaceship returns is difficult to fulfill. Perhaps this is why many students face it by doubting the possibility of moving "straight ahead" in the curved world of S^3. This is another interesting follow-up question (see next section).

In order for students to be able to construct the intended mental images and answer the mathematical questions, the following mental images are productive:

- It can be helpful to take into view only a part of the current world, imagining it locally. As the one-dimensional spaces $\mathbb{R}^1$ and S^1 allow only one degree of freedom for movement (forward/backward), students sometimes imagine them as tracks of a railway or of a roller coaster (looping).

- For the cases in which the local world is either one- or two-dimensional, students find it useful for the control and guidance of their mental images to have access to an outside perspective in addition to the inside one. If they imagine, for example, that the world constructed from the closed rope S^1 is suspended in the three-dimensional space $\mathbb{R}^3$, they can imagine the movement as a whole (such as how the circle of rope or the looping is traversed again and again). Students can take this outside view either through imagining they are zooming out or by changing their perspective.

- Students sometimes embed the one- and two-dimensional worlds in $\mathbb{R}^3$ and imagine them globally, as a whole. It is even more productive to imagine $\mathbb{R}^2$ lying in $\mathbb{R}^3$ and S^1 lying in S^2. Then, for example, the straight-ahead movement on the surface S^2 might follow the equator, which is a circle S^1. Students may find it helpful to track the movement on the sphere's surface by leaving a colored trail behind.

I now explain how it is possible to approach obtaining a mental image of the hypersphere in the following section.

Mathematical follow-up questions

Two questions are coming up mostly every time in my classroom: How can we imagine a hypersphere as a whole? And how can we move straight ahead on a surface that is not flat but curved as the sphere S^2? As the second question is easier to answer, let's start with that one.

After some discussion in class, it becomes clear that moving straight ahead means to take the shortest path, to follow a path that is as "noncurved" as possible: a path in S^2 is "straight" when it bends neither to the left nor to the right (see also R11).

Now, how could students imagine the hypersphere S^3 and the flight of their spaceship? In old geographical atlases, students can see the depiction of the Earth's surface as two circular disks, one for each hemisphere. Their circular boundaries are joined together like this: a point on the boundary of one disk is glued to the corresponding, mirror-inverted point on the boundary of the other disk. With these gluing rules, we obtain a purely two-dimensional representation of the sphere in two-dimensional space $\mathbb{R}^2$. Crossing the equator on the Earth's surface means crossing the boundary of one disk and entering the other disk at the corresponding point of its boundary. Note that the representation used here does not involve the third dimension. Analogously, S^3 can be understood as two solid balls in which pairs of boundary points are glued together (in Figure 8.1, two such gluings are shown.) Based on this analogy, the hypersphere can be represented in three-dimensional space $\mathbb{R}^3$ without reference to a fourth dimension.[2]

The flight in the closed three-dimensional space of the hypersphere S^3 emerges—analogous to sailing around the Earth following the two-hemisphere atlas described previously—as follows: Beginning from a position on the Earth,

---

2   For the topology of the hypersphere, see Weeks (2020, 193–206), and for its volume and surface area, see Coxeter (1973, 125–126).

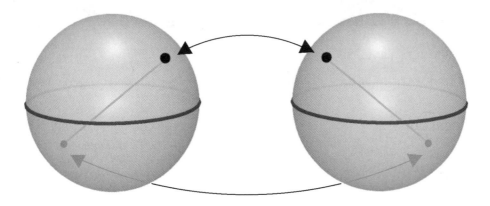

FIGURE 8.1 Two solid balls with surfaces glued together form a hypersphere

located within one of the solid balls, the traveler's path leads sooner or later to this solid ball's boundary. At this position, the path changes to the corresponding boundary point of the second solid ball and then goes on through its interior and leads once again to the boundary. The journey continues on the first ball and finally arrives back at the starting position—the Earth. From here, the journey can go on in any direction. Although the hypersphere is closed, there are no restrictions on the travel possibilities there.

For the sake of this imagining, let's ignore the finite speed of light and assume that light spreads in a straight line without any delay. Then, a universe with the topology of a hypersphere would have the following surprising characteristic: a ray of light emanating from the far side of the sun reaches an observer who is gazing into the sky on the far side of the Earth (with respect to the sun)—in other words, at night. As a result, we would see the sun as a tiny point of light in the nocturnal starry sky. A similar characterization: from Earth, we could use a very high-powered telescope to observe the Earth itself. Having chosen a direction to point the telescope, we will see the surface of the Earth that lies under our feet in the "antipodal" direction. This would work regardless of which direction we point the telescope!

This thought experiment encourages students to ponder the meaning mathematics has for our worldview. As the example of the hypersphere demonstrates, the global structure of the universe could actually be closed without limiting our freedom of movement in everyday life. Interestingly, one of the presently unanswered questions of cosmology is whether the universe is infinite or finite in extent. Based on today's experimental data, cosmologists cannot tell whether the universe is finite or infinite.

P3   SLINGING A BALL ATTACHED TO A STRAP

DIFFICULTY LEVEL: ★ ★

MATHEMATICAL IDEA: Whether and how an object moves depends also
on its reference system.

PREREQUISITES:
- If you spin around while holding a strap attached
to a ball, the ball is forced onto a circular
trajectory.

Whether we look at the moon with our naked eyes or through a telescope, it always shows us its same side. Can we conclude from this that the moon doesn't rotate around itself? The German astronomer Johannes Kepler is said to have been of this opinion: The sun rotates, he reasoned, to impart motion to its planets, and the Earth rotates to impart motion to its moon. Since the moon has no smaller moon of its own, there is no need for it to rotate.

This exercise in Mathematical Imagining disguises the question of lunar rotation in the context of a ball. Your students imagine the ball is on a strap, and they are swinging the ball around themselves. In doing so, they always see the same side of the ball while they are swinging it around.

- Imagine there is a *ball* lying in front of you, and it is attached to a *strap*. . . .

- Take the free end of the strap in your hand and begin to *spin*. Turn around and around so fast that the ball is brought into a circular trajectory.

- Is the ball also *rotating around itself* while you are spinning around—or not? On the one hand, the ball is rotating around itself because you are swinging it around. On the other hand, the ball is not rotating around itself because you always see the same side of it, namely, the side where the strap is fastened to the ball.

- Which answer do you choose? What arguments can you bring to the discussion that rebut the opposing answer?

- What did you imagine in this exercise in Mathematical Imagining?

Comments

The answer depends on the point of view. If you take an outside position and observe the person with the ball, the ball rotates around itself simply because it is going around the person. From the perspective of the person swinging the ball, it is not rotating, as the reference system—the person—is already turning. Relative to the person, the ball is still. The position and changes to it are relative and depend on the selected reference system.

Mathematical follow-up questions

If the moon (as viewed from the Earth) rotated, for example, one full turn for each orbit around the Earth, and like the Earth, counterclockwise when viewed from above, then when observed from a point outside the Earth–moon system, the moon would make two full turns. Analogously, you can pose the question, "How many full turns does a coin make if you roll its circumference—without slipping—along the circumference of another, same-sized coin?" Your students can explore this *coin rotation paradox* in class, too, within the framework of an imagining task. (See Chapter 3 for suggestions about writing your own imagining tasks.)

Based on the fact the both coins have circumferences that are the same size, it is tempting to assume that the rolling coin will make one full turn per rotation around the stationary coin. If your students give this answer, then they are focusing on how the circumference of the rolling coin rolls along that of the stationary coin, meaning they have placed their reference system at the point where both coins touch. From the local point of view (of the moving coin) of this reference system, the rolling coin does, in fact, make only one full turn. However, because this point of contact is moving along a circular trajectory, the local rotation of the coin combines with the rotation of the reference system. Since both rotations happen at the same time, an "outside" observer will see that the rolling coin makes two full turns ($720°$).

How many times does the coin rotate when it is rolled along the inner side of a circular-shaped ring with a diameter or a circumference that is twice the size of the coin's? In this case, the coin's circumference can be measured off twice along the circumference of the ring. This is why, when viewed from the local perspective of the point of contact, the coin makes two full turns. In contrast to the first case, this local rotation of the coin is combined with an opposite rotation of the reference system. The end effect is that, for the outside observer, these two rotations combine into a full turn of $360°$ contrary to the direction of rolling.

P4   COMPARING THE NUMBER OF POINTS ON AND THE LENGTHS OF CIRCUMFERENCES

DIFFICULTY LEVEL: ★ ★

MATHEMATICAL IDEA: The points of two circles can be matched one-to-one with each other even if the circles are not the same size.

PREREQUISITES:
- Turning one clock hand on two circular, concentrically arranged clockfaces of different sizes.
- If two curves consist of a finite and equal number of segments of equal length, they are also of equal length as a whole.

We tend to imagine a line as a series of points packed very tightly together. In doing so, we are essentially reproducing an atomism according to which all matter is composed of tiny particles. Euclid himself probably realized that this scientific concept could not automatically be applied to geometric objects because he conceived a line as "breadthless length."

In this exercise in Mathematical Imagining, two circles of different sizes are arranged concentrically. The infinitely many points of both circles are matched *bijectively* with each other by rays radiating from their common center. Each element of the first circle set corresponds with exactly one element of the second circle, and each element of the second circle corresponds with exactly one element of the first circle. The introduction of two clockfaces and a clock hand supports the plausibility of this correspondence. Each time mark on the inner circle corresponds precisely to one on the outer circle. There thus appear to be the "same number" of points on both circles, despite the fact that they are of differing lengths. As in task P1, the subject of this exercise in Mathematical Imagining is once again actual infinity. While P1 looked at the "calculation with infinite sets," this task is about the "size of infinite sets." In particular it's about the size of infinite sets that, unlike the rooms of Hilbert's Hotel (exercise P1), cannot be counted, that is, put into a one-to-one relation with the natural numbers.

- Imagine two *clockfaces* lying in front of you on your desk. Both clockfaces are *circular,* but they do not have the same size. . . .

- Pick up the smaller clockface and place it on top of the larger one so the two clockfaces are on top of each other. Line up the circle centers so they are *directly above each other*. . . .

- Now look at the common *center* of the two clockfaces. Attach a *long* and *thin* clock hand to this point so it extends beyond the larger clockface. In doing so, it first intersects the boundary of the *inner* clockface and then the boundary of the *second*, outer clockface. . . . To be more precise: the clock hand intersects the inner circle at *one* point and also the outer circle at exactly *one* point. . . .

- Turn the clock hand forward a little bit. . . . No matter how far you have turned the hand, it will connect *one point on the periphery* of the inner clockface with *exactly one point on the periphery* of the outer clockface.

- In this way, *each point* of the inner circle corresponds exactly to *one point* of the outer circle, and, conversely, *each point* of the outer circle corresponds exactly to *one point* of the inner circle. . . .

- Does this mean that the corresponding circumferences also have the same *length*?

- What did you imagine during the exercise in Mathematical Imagining?

Comments

The mathematical question at the end of the imagining task makes the contradiction inevitable. The statement that the points of two geometric objects can be matched one-to-one with each other does not, in the case of infinite sets of points, imply that the two objects must also be of the same length. While the geometric length of a set is a matter of its *measure*, the number of elements of a set is a matter of its *cardinality*—two very different concepts that can be used to determine the "size" of a set.

With the concept of cardinality, the concept of the number of elements is generalized to infinite sets. As soon as two finite sets have the same number of elements (e.g., seventeen), they are *equipotent*. Infinite sets can no longer be compared by counting. To establish the equipotency of two infinite sets, one therefore creates pairs of elements from the two sets. As each point on the inner circumference corresponds by means of a ray to exactly one point on the outer circumference, the two circumferences are equipotent—despite their differing lengths. The concept of cardinality resolves the contradiction that seems to be inherent in the statement "lines of different lengths contain the same number of points."

Mathematical follow-up questions

Since this result does not depend on the radius of the particular circle, all circles are equipotent, regardless of their length. In the same way, every line segment is equipotent to any one of its subsegments, however short. Infinite sets are characterized by the fact that they have (at least) one proper subset that is equipotent to the whole set.

We have a similarly paradoxical situation when one large and one small wheel are attached to a single axis, sometimes called *Aristotle's wheel paradox*. If the large wheel is rolled once along a line, the small wheel also turns exactly one time. Does this mean that it has "rolled" along a segment of the same length as the large wheel? No, of course not, because the small wheel doesn't only roll, it also slides. On every section of the line on which the small wheel rolls, however small, there are points at which it both rolls and slides.

From a set theory point of view, then, sets can be indistinguishable even when they differ geometrically. This applies not only to length but also to geometric dimensions. For instance, the points of a line can be matched bijectively with the points of a square. Consequently, even squares (or cubes) do not have a cardinality greater than that of lines, despite the differing geometrical dimensions of the two sets.

P5   TURNING THE UNIVERSE INSIDE OUT

DIFFICULTY LEVEL: ★ ★ ★

MATHEMATICAL IDEA: The universe might be made in such a way that the length of an object changes when the object moves.

PREREQUISITES:
- Interior surface area and interior volume of a large hollow sphere.

- The series $\frac{1}{2} + \frac{1}{4} + \frac{1}{8} + \dots$ converges to 1 (see R2).

Physical objects keep their length when we transport them to a different location—at least this seem to be the generally accepted truth. However, this imagining task shows that things might behave completely differently. As in task P2, here again students will mentally experiment and question the conventional view that we have of the world around us.

This exercise in Mathematical Imagining describes a universe constructed on the interior surface of a large, but finite, sphere. The familiar surface of the Earth is exchanged for the sphere's interior surface, and even the planets and stars are inside this hollow sphere. In order to imagine such a universe so that it doesn't have contradictions, other physical laws have to apply. Here, the closer an object gets to the center of this universe (the center of the surrounding sphere), the shorter the object's length becomes. At the same time, an object's speed—and in particular the speed of light—lessens the closer it gets the center. To the inhabitants of such a universe, the lengths and speeds of moving objects appear constant. For outside observers, though, the objects are shrinking and slowing down. How can this contradiction be resolved?

- Imagine a *hollow sphere* with, let's say, a diameter of many miles. . . . Imagine that you—and all the other inhabitants of the Earth—live on the *interior surface* of this sphere, *cradled*, so to speak, within the sphere's curving. . . .

- This hollow world is made in a very peculiar way: as soon as you *rise up* from the sphere's interior surface, you start getting *smaller*. . . .

- Indeed, all objects start *shrinking* as soon as they move toward the *hollow sphere's center*. By the time an object is halfway to the sphere's center, its size has been *halved*. When objects keep getting closer to the center of this universe, they *shrink* even more. Every time an object travels half of the remaining distance to the center, it shrinks again to half its size. . . . Thanks to this property, though, you can also fit the sun, moon, and all the stars comfortably into this universe inside your hollow sphere. . . .

- Additionally, imagine that when objects move toward the center, their *speed* also slows down. As soon as objects are halfway to the center, their travel speed has been *halved*. . . .

- For you—as an inhabitant of the hollow, spherical world—your universe is infinitely big. No matter how "far" you fly toward the center, you can keep traveling farther and farther, without ever returning to the place where you started your journey. . . .

- Let's put aside the question of whether your hollow, spherical world is possible and concentrate instead on the finiteness or infiniteness of your universe. Imagine that, one day, a woman and a man that live in your hollow sphere are having a big debate because. . . .

- The woman sees her world as an *infinite universe* where all objects that move *keep* their size. . . . The man, though, sees his world as a *finite universe* where the size of objects *changes* when they move. . . .

- What arguments would you put forth to try and mediate the disagreement between these two people?

- What did you imagine in this exercise in Mathematical Imagining?

Comments

The decisive stipulation for the hollow, spherical world is that objects' length and speed reduce when they approach one particular point (the sphere's center). This is rooted in the geometric series $\sum_{n=1}^{\infty} \frac{1}{2^n}$. From point of view of the summation (grasped from "inside"), you can continue an operation like $\frac{1}{2} + \frac{1}{4} + \frac{1}{8} + \dots$ an infinite number of times without ever reaching the end in finite time. If you take the "outside" view and look at the sum—here it's 1—you can see that this operation does not go beyond the finite value, even when it is carried out infinitely many times. Indeed, the longer you perform the operation, the closer you get to 1. (See also task R2.)

Analogous to the partial justification of the two perspectives on the geometric series, the two inhabitants who disagree about whether their world is infinite or finite each have several mathematical arguments to support their respective positions. From the mathematical point of view, both world models are thinkable.

Seen in terms of physics, the hollow-sphere world breaks with the principle that the laws of nature are the same throughout the universe. Logically, this is not a problem, as this principle is a useful assumption. To date, the hollow-sphere world theory has not appeared to violate any of the classic, applicable laws of cosmology. In other words, both geometric models of the universe are thinkable from the physical perspective.

Therefore, this imagining task only *seems* to be contradictory. The conventional model just seems more natural to us because we tend to see ourselves as the navel of the world and prefer simple answers. The history of science, however, shows that this tendency has repeatedly hampered us in expanding traditional worldviews. Scientists must be prepared to encounter issues for which multiple, equivalent (geometric) models are thinkable, without being able to decide which model reflects (physical) reality. There are scientific findings that are more based on convention than actually reflecting the nature of the object of study.

Mathematical follow-up questions

Viewed geometrically, there is an *inversion in a sphere* within the cosmological model presented here. This transformation of a point P to a point P' can be described in two ways.

- Join the center M of the sphere and any point P (other than M) with a straight line. If d denotes the distance from P to M and the radius of the sphere is 1, then the image point P' is on the joining line of P and M at a distance $d' = \frac{1}{d}$ from M.

- Constructively, this means choose a plane passing through P and M. Letting this plane play the role of the equatorial plane, identify two points on the sphere, its North Pole N and South Pole S. Then connect N and P with a straight line; let K be the point where this line intersects the sphere. Finally, connect K to S; where this second line intersects the equatorial plane is the sought-for point P'.

This inversion in the sphere not only maps the sphere's outer surface onto its inner surface but also maps the entire exterior space onto the interior space (and vice versa, with the exception of the center M). With this, you can discuss with your students such questions as, How does the inversion in a sphere act on straight lines or planes, or on circles or spheres? How would light spread out in the hollow world? What trajectories would geocentric satellites take?

REFERENCES

All the Internet addresses here and in the book were last verified at the end of November 2019.

Begehr, H., and H. Lenz. 1998. "Jacob Steiner and Synthetic Geometry." In *Mathematics in Berlin*, ed. H. Begehr, H. Koch, J. Kramer, N. Schappacher, and E. J. Thiele (49–54). Basel: Birkhäuser.

Conway, J., P. Doyle, J. Gilman, and W. Thurston. 1991. "Exercises in Imagining." In *Geometry and the Imagination* in Minneapolis, version 2.0 (2018), 35–37. https://arxiv.org/pdf/1804.03055.pdf

Courant, R., and H. Robbins. 1996. *What Is Mathematics? An Elementary Approach to Ideas and Methods.* 2nd ed. New York: Oxford University Press.

Coxeter, H. S. M. 1973. *Regular Polytopes.* New York: Dover.

Cromwell, P. 1997. *Polyhedra.* Cambridge: Cambridge University Press.

Dewey, J. 1897. "My Pedagogic Creed." *School Journal* 54 (1): 77–80.

Engel, A. 1998. *Problem-Solving Strategies.* New York: Springer.

Eves, H. W. 1969. *In Mathematical Circles: A Selection of Mathematical Stories and Anecdotes.* Boston: Prindle, Weber & Schmidt.

Feynman, R. P. 1989. *What Do You Care What Other People Think? Further Adventures of a Curious Character.* New York: Bantam Books.

Freudenthal, H. 1978. *Weeding and Sowing: Preface to a Science of Mathematical Education.* Kluwer Academics.

Gallin, P., and U. Ruf. 1998. "Furthering Knowledge and Linguistic Competence: Learning with Kernel Ideas and Journals." *Vinculum* 35 (2): 4–10. http://www.ecswe.org/wren/documents/Article1GallinVinculum.pdf.

Galton, F. 1883. *Inquiries into Human Faculty and Its Development.* London: Macmillan. http://galton.org/books/human-faculty/.

Glaeser, G., and K. Polthier. 2020 (in press). *A Mathematical Picture Book*. Berlin: Springer.

Hilbert, D., and S. Cohn-Vossen. 1952. *Geometry and the Imagination*. New York: Chelsea.

Isenberg, C. 1992. *The Science of Soap Films and Soap Bubbles*. New York: Dover.

Johnston-Wilder, S., and J. Mason. 2005. *Developing Thinking in Geometry*. London: Paul Chapman.

Kawohl, B., and C. Weber. 2011. "Meissner's Mysterious Bodies." *Mathematical Intelligencer* 33 (3): 94–101.

Mason, J. (2002). "Exploiting Mental Imagery in Teaching and Learning Mathematics." *Actas do ProfMat 2002*, 75–81.

Nelsen, R. 1993. *Proofs Without Words—Exercises in Visual Thinking*. Washington, DC: Mathematical Association of America.

Petit, J.-P. 1986. *Here's Looking at Euclid (and Not Looking at Euclid)*. Los Altos, CA: William Kaufmann. http://www.savoir-sans-frontieres.com/JPP/telechargeables/English/HERE_S_LOOKING_AT_EUCLID.pdf.

Pólya, G. 1945. *How to Solve It.* Princeton, NJ: Princeton University Press.

———. 1962. *Mathematical Discovery: On Understanding, Learning, and Teaching Problem Solving*. Vol. 1. New York: Wiley & Sons.

Thom, R. 1973. "Modern Mathematics: Does It Exist?" In *Developments in Mathematical Education*, ed. A. G. Howson (194–209). Cambridge: Cambridge University Press.

Wagenschein, M. 1970. *Ursprüngliches Verstehen und exaktes Denken, I–II* [Original Understanding and Exact Thinking, vols. 1–2]. Stuttgart, Germany: Klett Verlag.

Weeks, J. R. 2020. *The Shape of Space*. 3rd ed. Boca Raton, FL: CRC Press.

Weisstein, E. 2003. *The CRC Concise Encyclopedia of Mathematics*. 2nd ed. Boca Raton, FL: CRC Press. (For updated entries, see Wolfram MathWorld, http://mathworld.wolfram.com/.)

Wertheimer, M. 1959. *Productive Thinking*. New York: Harper & Brothers.

INDEX

(Note: Page numbers followed by *n* refer to footnotes; page numbers followed by *t* refer to tables.)